NATIVE AMERICANS
INTERDISCIPLINARY PERSPECTIVES

Edited by
John R. Wunder
Cynthia Willis Esqueda
University of Nebraska–Lincoln

A ROUTLEDGE SERIES

Blood Matters

The Five Civilized Tribes and the Search for
Unity in the Twentieth Century

Erik M. Zissu

Routledge
New York & London

Published in 2001 by
Routledge
29 West 35th Street
New York, NY 10001

Published in Great Britain by
Routledge
11 New Fetter Lane
London EC4P 4EE

Routledge is an imprint of the Taylor & Francis Group
Printed in the United States of America on acid-free paper

Copyright © 2001 by Erik M. Zissu

All rights reserved. No part of this book may be reprinted or reproduced or utilized in any form or by any electronic, mechanical, or other means, now known or hereafter invented, including photocopying and recording, or in any information storage or retrieval system, without permission in writing from the publisher.

10 9 8 7 6 5 4 3 2 1

Library of Congress Cataloging-in-Publication Data
Zissu, Erik M., 1965–.
　Blood matters : the five civilized tribes and the search for unity in the early twentieth century / by Erik M. Zissu.
　　p. cm.
　Includes bibliographical references and index
　ISBN 0-415-93086-3
　　1. Five Civilized Tribes—Ethnic identity. 2. Five Civilized Tribes—History—20th century. 3. Indians of North America—Indian Territory—Ethnic identity. 4. Indians of North America—Indian Territory—History—20th century. I. Title.
E99.I5 Z57 2001
976.6'00497—dc21   2001031907

Now, the Star-Belly Sneetches
Had bellies with stars.
The Plain-Belly Sneetches
Had none upon thars.

Those stars weren't so big. They were really so small
You might think such a thing wouldn't matter at all.

Dr. Seuss
The Sneetches

"Oh," Christmas said. "They might have done that? dug them up after they were already killed, dead? Just when do men that have different blood in them stop hating one another?"

William Faulkner
Light in August

Contents

PREFACE	ix
A NOTE ON TERMS	xiii
ONE Introduction: "A Disgrace to My Blood"	3
TWO Lament for a Golden Age	9
THREE "Real" Indians	23
FOUR The Importance of Being White	37
FIVE Hardship and Decline	51
SIX Assimilation's Failure	71
SEVEN "Indians, Organize!" A Peoplehood in Blood	87
EIGHT Epilogue	109
ENDNOTES	111
BIBLIOGRAPHY	141
INDEX	149

Preface

BLOOD MATTERS. IT DEFINES WHO YOU ARE; IT FURNISHES AN UNALTERABLE HISTOry and an ineradicable orbit of relationships. Within the wider society, blood serves to define and categorize. Certain peoples are discriminated against on the basis of their racial ancestry. At the same time, ancestry is employed to defend group distinctiveness. Blood is held up with pride or denied as taint.

Consequently, blood has come to occupy a central place in national discussions of identity politics. Some practitioners of race, whether out of meanspiritedness or the belief in an objective science, dote on biology in order to discern character and behavior. This insistence upon marked racial difference is evident in battles involving education, crime, and countless other areas of American life. It has also provoked a dedicated response. There have been over the past decades repeated calls to ignore, to look beyond, blood. Race, it is maintained, is not biological but rather artificial; it is manufactured by society. Yet despite rational arguments and eloquent appeals, blood remains a stubborn element in group relations. However frequently it is used to mask prejudice, or is explained as a "social construction," it is potent. Bloodstains resist removal.

During the research for this study, I was struck many times by references in historical documents to blood. Blood repeatedly appeared as a form of shorthand for identity; character and personality were attributed to it. It also conveyed significant material power. Blood might dictate opportunities available to an individual or, conversely, limit an entire group. But while the papers I sifted through in various archives in Oklahoma, Arkansas, Texas, and Washington, DC, were saturated with references to blood, it was only by coming across contemporary examples of individuals pursuing familial blood trails that I began to sense a theme in the history I was studying. Two anecdotes begin to illustrate this.

I recall conducting research one day in the Oklahoma Historical Society when I left my laptop computer and notepads in an empty research room while I went for lunch. When I returned, I noticed that the parking lot had filled to capacity and, upon re-entering the research room, that my belongings had been moved into a corner to accommodate the numerous people who were now searching through reels of microfilm. Many, I noticed, were elderly. Some had their children and grandchildren in tow. Judging by the license plates on their recreational vehicles outside, these people had traveled substantial distances.

I asked the director of the archives, William Welge, who these people were and what they were doing. His response was simple: "They have come to prove they are Indian." They scoured old census lists to track down the slightest connection to an Indian ancestor, often looking to prove a lineage that had been but a whispered family rumor, a tantalizing allusion to a special legacy. These folks did not live on reservations, and they seemed to possess no trace of Indian culture or knowledge. But they doggedly hunted down suspected ancestors, traveling from California, New York, and other distant places to wade through the hard-to-read records of long dead people.

Welge told me that many people came to Oklahoma City because they thought that securing an Indian ancestry would bring them some form of material advantage: free health care, per capita payments generated by oil leases or bingo hall profits, or other hoped-for windfalls. Blood, in a very real sense, might yield actual benefits. But other, less material motivations drove these genealogical researchers. Laying claim to an exotic heritage seemed to hold out personal fulfillment. Within the mundane and mind-numbing census lists, attached to distant, unknown relatives, a unique birthright beckoned.

Perhaps more surprising, and in terms of this study more piquant, was an encounter I had with a Seminole in the same research room. After completing a portion of my own work one day, I was introduced to a visibly elated man who appeared to be in his late 50s or early 60s. When I asked why he was excited, he responded that his genealogical digging had turned up some good news. He had just discovered a new branch on his family tree. "I am now," he said, "five-sixths Seminole. One more and I'm a full blood."

This man was unmistakably Indian in appearance, and he was an enrolled member of his tribe with all the benefits, and liabilities, that came with official membership. Why then, I wondered, would a designation of full blood be important? I could understand desiring a full accounting of one's ancestry, but not of an effort to upgrade oneself from four-sixths to five-sixths to full blood. As a member of the Seminole nation, his material benefits would be unaffected by blood quantum. Perhaps, I reckoned, he had other, more distant concerns on his mind. How might his research affect his children? His grandchildren? Blood, after all, extends down through the generations affecting people in various ways. A full blood might ensure that later generations would possess stronger claims to membership. I did not press him as to why being full blood was so significant, for I believed that the matter was somehow too personal for a stranger to delve into. But I was reminded of the importance of blood. The discovery of Indian ancestors even for, perhaps especially for, an Indian had import beyond the research room.

My hesitancy in asking why blood mattered to the Seminole man I met in Oklahoma suggests an important consideration. Race is a sensitive issue and often politically explosive. In writing about these topics, I do not advocate biological determinism. Rather, my interest lies in the uses of race, especially those of a political nature. As the stories of the research room at the Oklahoma Historical Society suggest, there are widespread applications of race by individuals, as well as groups, in American society.

Although there have been efforts to disassociate Indian politics from race, to focus, for example, on historic treaty rights and the sovereignty implied in those rights, there can, I believe, be no simple disengagement of race and politics. In a recent *New York Times* article, a California Indian leader, addressing current efforts to obtain sovereignty for recognized tribes, was quoted as saying: "What most people don't understand is that we are governments first, and racial entities second." But those same governments rely strongly, if not solely, upon a single criterion in determining their membership: blood. Matters of ancestry and politics are knotted and cannot be simply untangled.

Let the above anecdotes serve as preface. While the answers to many of the questions raised here exceed the reach of the coming pages, I aim to provide a bit of historical groundwork for contemplating the ways people conceive of and use blood. Although it is frequently employed with cruel intent and can serve as a damning indictment of whole groups, blood is a versatile tool. For Indian groups, specifically the Five Civilized Tribes in the wake of Oklahoma statehood in 1907, blood has served many purposes.

* * *

I wish to acknowledge those institutions and individuals rendering assistance in the preparation of this study. Generous and necessary research and travel grants were provided by the Phillips Fund for Native American Studies of the American Philosophical Society and the Redd Center for Western Studies at Brigham Young University. Two Andrew W. Mellon Predoctoral Fellowships from the University of Pittsburgh gave me the opportunity to finish my research and to write. In addition, the University of Pittsburgh Department of History extended teaching-assistant and teaching-fellow scholarships that made preliminary study possible, including the successful completion of the Master's degree, the comprehensive Ph.D. examinations, and the dissertation overview.

During the course of my research, I was aided by the staffs of various facilities, including the National Archives in Washington, DC, the Federal

Archives and Records Center Southwest Region in Fort Worth, Texas, the Oklahoma Historical Society in Oklahoma City, the Western Historical Collection and the Carl Albert Congressional Research and Studies Center Congressional Archives at the University of Oklahoma, and Hillman Library at the University of Pittsburgh. In addition, Prof. Daniel F. Littlefield, Jr., allowed me access to the American Native Press Archives at the University of Arkansas at Little Rock of which he is the director.

A number of individuals assisted me in the course of this study's evolution. Foremost among those offering encouragement and guidance was Prof. Robert Doherty, my advisor. Simply stated, without his patience, advice, and constant effort to push me to develop my ideas, the manuscript would never have materialized. The commitment Prof. Doherty made to me and this project extend well beyond the usual scope of a mentor's obligation, and to him I owe an incalculable debt. I am most glad that he has become my friend, as well as my teacher, during my graduate studies. Michael Naragon lent me his ear and his time, both in unselfish portions. Good friend that he is, he listened, and he listened, and he listened. When I let him, he interjected, and always for the better. He also provided detailed, insightful comments on the entire project. Patrick Dowd and Leslie Hammond, perhaps unfairly twinned, also endured me and this project over the years. Among other lessons, they both taught me the importance of questioning my assumptions. Their friendship has made pursuing the doctorate a near bearable experience. Prof. Donald L. Parman of Purdue University has taken the time to assist me in various capacities over the past several years. He hosted me in West Lafayette and provided me with timely feedback and advice. He also allowed me to pick his brain while generously buying me meals in such places as Indiana, Oklahoma, and Minnesota. Prof. Daniel Littlefield has offered his expertise in matters of Five Tribes history and his enthusiasm in helping me to explore some of the ideas in the dissertation. Professors Paula Baker, Marcus Rediker, and David Brumble were kind enough to serve on my dissertation committee and supply timely and generous feedback. They entertained my ramblings at various stages of this study's development and waited patiently for the finished product.

Finally, my thanks go to my family. Everyone stood back at a respectful distance and periodically asked me when the work would be finished. My mother and Peter learned to accept, graciously, "soon" or "a few years" as a satisfactory response, as did my father. Happily my family has expanded in recent years, and while Zoë and Max live in blessed ignorance of this study, Kris is all too aware of its existence. To her the book is therefore dedicated.

A Note on Terms

UNFORTUNATELY, THE IMPERFECTION OF LANGUAGE EXISTS ALONGSIDE ITS CAPACity for insight. In this study, certain terms are used, some with great frequency, to identify groupings of members within the Five Civilized Tribes of eastern Oklahoma. Within these tribes—the Cherokees, Chickasaws, Choctaws, Creeks, and Seminoles—there existed various factions. Scholars investigating the development of the tribes, and indeed all Indian groups, have tried to apply labels to factions, and the results have been uneven. There were full bloods and mixed bloods, conservatives (or traditionalists) and progressives, small and subsistence farmers and large landholders, separatists and integrationists. While I approach these terms with an awareness of their liabilities, I nonetheless have used them all at various points in the coming pages. For the most part, I rely on the terms conservative and progressive to convey a general impression about the two largest, most encompassing groupings within the tribes. Conservatives tended to shy from white society and preferred lives lived among their tribal fellows where older traditions and practices might be preserved. They more likely possessed significant levels of Indian ancestry, a fact that they were well aware of and used for various purposes. Progressives displayed greater familiarity with, and attraction to, white values. They moved more easily within white society and embraced economic development of Indian Territory and, subsequently, the state of Oklahoma. Progressives, almost exclusively, derived from mixed whiteIndian parentage and not infrequently had only the scantest biological ties to their tribes.

While I recognize the inadequacy of these terms, they nonetheless provide a measure of continuity with previous historical writings on the Five

Tribes. They also lend a sense of how I view the widening gulf that separated tribal members from one another as they endured the substantial changes sweeping over their homes during the late nineteenth and early twentieth centuries. At the same time, however, it is my hope that this study forces the reader to consider the complexity of issues of identity among tribal members.

Blood Matters

ONE

Introduction: "A Disgrace to My Blood"

Nearly a hundred years after it was "hurriedly" written aboard a steamship heading west from Mississippi, a letter by the Choctaw leader George Harkins was re-published in a 1926 essay by the Choctaw historian Muriel Hazel Wright. Harkins addressed his missive "to the American people," summing up the struggles of his fellow tribesmen as they contended with removal from their homelands in the Southeast and the "mountain of prejudice that has ever obstructed the streams of justice." "We found ourselves like a benighted stranger," Harkins wrote of the challenges of recent years, "following false guides, until . . . surrounded on every side." Given only the cruelest choices for survival, the tribe opted for freedom, the noblest goal of any people: "We as Choctaws chose to suffer and be free, than live under the degrading influence of laws, where our voice could not be heard in their formation."[1]

Harkins' letter struck a chord in Muriel Wright, who later went on to edit the *Chronicles of Oklahoma* and publish several seminal works on Five Tribes history. Knowing the consequences of the Choctaw removal, the perils of winter travel undertaken by a weakened and ill-clothed population, she admired the chief's tenacity, his stoic acceptance of duty, and the rightness of his actions. She was impressed with his leadership and his obvious ability to discern and give voice to "the feelings of his people." As an historian, Wright also could not help but to see the parallels that Harkins drew between the Choctaws and a certain other collection of people that had faced hardship in establishing a nation in "a wild country that was open to many enemies including . . . thousands of hostile Indians."[2] Had not the American colonists, too, chosen to "suffer and be free" rather "than live under the degrading influence of laws, where," as Harkins explained, their "voice could not be heard in their formation"?

The purpose of Wright's essay, the centerpiece of which was Harkins' letter, might well have been attained with the comparison of the Choctaws' plight with that of the United States itself. The irony of Indian mistreatment at the hands of freedom-loving Americans, of the tyranny spawned by democratic ideals, neatly argued for a re-assessment of the nation's history. It suggested that a careful look at the Indian past might yield new insight into peoples generally dismissed as peripheral to the country's development. But the Choctaw chief's letter had a further relevance. In the early twentieth century, far from having overcome adversity, the Choctaws continued to face immense difficulties, not the least of which was the federal government's insistence that Indians forsake their tribal associations in favor of adapting to white society and the pursuit of individual success. In the face of these continued challenges and the toll they had taken upon her tribe, Wright appreciated Harkins' insistence on tribal unity. While Harkins acknowledged that individual Indians, himself included, might find a measure of success in white society, the cost was too great. "I must go with them; my destiny is cast among the Choctaw people," the chief had written of his own exile westward. "If they suffer, so will I; if they prosper, then I will rejoice. . . . Could I stay and forget them and leave them to struggle alone, unaided, unfriended? . . . I should then be unworthy of the name of Choctaw, and be a disgrace to my blood."[3]

The struggle for unity, which provides a focus for this study, is a misunderstood aspect of the history of the Five Tribes in the era of Indian Territory's dissolution and the birth of the State of Oklahoma. The predominant story of the Five Tribes during this period generally emphasizes the disorganization and demoralization suffered by tribal members. The locus, understandably and undeniably, is the loss of a material resource base and the attendant downward trajectory of Indian culture and life. The best known writer of Five Tribes history, Angie Debo, crafted a portrait of victimization that bores deep into the consciousness of any and all who endeavor to learn about the tribes and their efforts to survive. Exhaustively, she documented the wholesale looting of tribal lands and resources, a process that resulted in, among other developments, bitter tribal squabbling. Under the crushing weight of American society, members of the tribes grimly fought against their exploitation, bearing witness to the crimes committed against them. But to end there, to leave the Five Tribes on the "Road to Disappearance," as the title of one of Debo's books suggests, is misleading.[4]

This book neither downplays nor obscures the losses suffered by members of the Five Tribes. Indeed, it employs a context in which Indians were discriminated against, robbed, and subjected to dispossession by federal and state policies. Yet within this context, tribal members did not relinquish the political initiative; they were not reduced to helplessness. Instead, they undertook political activity and fashioned a renewal of their collective identity that belied their apparent powerlessness. Although there arose a period of tribal disintegration following the dismantling of sovereign tribal governments and the division of commonly held lands, this was followed by an innovative, multi-faceted campaign waged by diverse groups within the tribes to bring about agreement and unified action. In grass-roots movements and intertribal political conventions and through an array of writings, tribal members resourcefully borrowed ideas and strategies from mainstream society and turned them to advantage. In particular, they used blood—as an integral component of a larger notion of Indianness—as a foundation for solidarity. A contrivance foisted upon Indians to promote assimilation, as well as a burdensome mark of inferiority, blood became a powerful tool in the hands of tribal members.

Importantly, these developments derived from the efforts of different "types" of Five Tribes members, men and women of widely varying background. While the diversity of tribal populations—within which there existed numerous distinctions of class and culture that derived from the tribes' lengthy exposure to non-Indian peoples—contributed to turmoil following the end of tribal sovereignty, it also enhanced the range and creativity of Indian response to life in early twentieth-century Oklahoma. As they were forced to define anew their Indianness within the increasingly hostile atmosphere of the region, tribal members found support from one another. A vital aspect of this study is the investigation of those tribal members generally referred to as "progressives." Most often of mixed, white-Indian parentage, these individuals have tended to be regarded as less Indian than their "conservative" and full blood fellow tribesmen.[5] Dismissed as assimilationists, they seemed to embody the larger changes sweeping the region, their own hybrid lineage pointing toward the disappearance of Indians. But progressives exhibited a strong sense of responsibility to their fellow tribesmen, and they deserve to be returned to a prominent place in tribal history. Indeed, their efforts to preserve tribal interests and to promote a notion of Indian peoplehood are central to the story told here.

Muriel Wright, well-educated and successful, was among those who recognized the hazards of internal disunity. She understood that earlier divisions among the tribes, divisions that paralyzed the Indians at the end of the nineteenth century, barred tribal members from effectively confronting non-Indian settlement and federal authority. Echoing the advice of George Harkins, she celebrated her Indian heritage and rejected government efforts at detribalization. Those agents of American civilization who demanded that Indians abandon their tribal existence were, as Harkins had labeled them, "false guides." During a period of unrelenting assault upon Indians and their remaining material assets, Wright and other members of the Five Tribes sought to regain control over their lives.

"Blood Matters" proceeds in a roughly chronological order. Chapter Two, drawn from accounts by other scholars, sketches the diversity of the Five Tribes in the late nineteenth century and emphasizes the ability of the autonomous tribal nations to contain internal conflict. The varied interests of tribal members co-existed harmoniously, albeit with occasional disruptions, within the independent nations of Indian Territory. But when external pressures grew too great, the tribes fractured. In the wake of the Dawes Commission's arrival in the territory, the polarization of tribal groups hardened, and Chapter Three documents how separatist-minded tribal members, for the most part small and subsistence farmers who wished to distance themselves from whites, accused their fellow tribesmen of betrayal. In so doing, they came to regard themselves as "real" Indians and their opponents as whites. One of the more significant criteria used for legitimating themselves as "real" was blood quantum.

Chapter Four explores how integrationist-minded tribal members, large landholders and those acquainted with and comfortable in white society, also employed blood to distinguish themselves among the tribal populations. Pursuing a course of assimilation in which resistance was fruitless, if not counterproductive, these tribal members emphasized their mixed ancestry in order to demonstrate their qualifications for equality with whites. As they moved away from full bloods, they also found it necessary to disassociate Indians from blacks, both actually and in the public's perception. In so doing, they displayed a nuanced grasp of the power of racial politics and attempted to ensure their continued access to opportunities.

The decline of the Five Tribes in the wake of Oklahoma statehood is outlined in Chapter Five. Struggling to survive in a place being re-shaped by powerful external economic forces, tribal members were reduced to the margins of society and extreme poverty. Indian powerlessness was further heightened by the rising confidence of the state government and the criminal acts of whites, who swindled tribal members relatively free from prosecution by the authorities and from feelings of guilt imposed by their own consciences. In response to these developments, progressives acted to lessen the dislocation of their fellow tribesmen. Operating through the Indian bureau, they recognized an obligation to those tribal members less prepared for the challenges of life in Oklahoma. Although they did not themselves feel threatened, they worked within the confines of government to improve basic material conditions.

Chapter Six traces the broadening of the threat facing the Five Tribes. The federal government, conceived as the best buffer between the tribes and white Oklahomans, appeared to be withdrawing its commitment to Indian matters. In addition, a rising level of racism and Indian victimization stemmed from the upheaval accompanying the boom of the oil industry. Resentment of oil-rich Indians spurred whites to vigorously plunder tribal members. Progressive tribal members discovered that they no longer were free from the bigotry and discrimination that had generally been reserved for more "backward" Indians. In response, they began to reach out to their fellow tribesmen and to formulate ideas about Indian solidarity. Recognizing

Introduction: "A Disgrace to My Blood"

anew their Indianness, progressives sought to bolster the tribes' distinctiveness in white society.

With Chapter Seven, the study comes full circle. Both progressives and conservatives, recognizing that their ineffectual responses to negative developments in Oklahoma derived, in part, from internal factionalism, engaged in a variety of political actions directed toward unity. New organizations were formed, large conventions were held, and grass-roots movements flourished. Importantly, diverse tribal members—ranging from separatists to a literary elite that formulated elaborate notions of Indian racial superiority and rewrote tribal histories in an effort to demonstrate the civilization of their forebears—utilized blood to fashion inclusivity. Whereas they previously employed ancestry to divide themselves during and immediately following allotment and the dismantling of their sovereign nations, tribal members by the 1920s considered blood a crucial bond. Accordingly, groups argued for unity regardless of blood quantum, stressing instead the mere existence of shared bloodlines. Tribal authors revealed the extent of this process by suggesting that the smallest fractions of Indian blood were enough to transmit the legacy of long dead "noble" ancestors.

TWO
Lament for a Golden Age

O N THE EVE OF OKLAHOMA STATEHOOD, THE CREEK LEADER PLEASANT PORTER appeared before a Senate committee that was touring Indian Territory. He used the occasion to lament the reduced state of the Five Tribes, placing the period in sharp contrast to an earlier era of tribal independence and prosperity. Referring to the years preceding allotment, a period seemingly more distant than the decade that had elapsed since surveyors began allocating individual homesteads to tribal members, Porter described a golden age when Indians "always raised enough to eat, and that was all we wanted." "[I]n my early life I don't know that I ever knew of an Indian family that were paupers," the aging chief told the assembled Senators, who had come to hear testimony as to the Indians' prospects as citizens in the nation's forty-sixth state. "They were all prosperous and happy and contented in their way and what more could they want?"[1]

In the past, Porter added,

> [w]e had a kind of Arcadian government. . . . If anyone was sick or unable to work, the neighbors came in and planted his crop, and they took care of it—saw the fences were all right—and the women took care of the garden, and wood was got for him, and so on. In fact, everything was done under the care of the people—they did everything and looked after the welfare of everything. . . . We have striven in our own way for our elevation and uplifting, and for a time it seemed that we were actually going to evolve a sort of civilization that would suit our temperament.[2]

Pleasant Porter's wistful recollections were colored by pride in past achievements and anguish at their ruin. His people had lived well; they provided for themselves and controlled their own lives. There was strength of community evident in Porter's backward-looking glance, the type born of hardship and renewal. But while his assessment touched upon certain truths, at the same time he omitted other, central characteristics of tribal life. The populations of the Five Tribes displayed a remarkable economic, cultural, and social diversity that far exceeded Porter's humble, one-dimensional portrait. Moreover, this diversity produced unavoidable conflicts. Throughout the late nineteenth century, tribal members continually differed in their views of a variety of issues, none greater than on how best to respond to the steady movement of American society westward.

Yet, despite his tendency to overlook the complexity of tribal relations, Pleasant Porter's sympathetic interpretation of the past retained a basic accuracy. Indeed, perhaps the most notable achievement of the tribes had been their ability to shape conflicts and foster compromise among competing interests. Tribal institutions channeled dissension toward discussion and debate; they allowed for opposing views to co-exist. While the rising presence of white settlers, ranchers, and railroad companies produced diverse responses from tribal members, ranging from defiance to collaboration, time and again the tribes reconciled internal disputes. In large part, the Five Tribes were aided in this delicate task by the fact of their sovereign authority over abundant material resources. They exercised true independence. This, in turn, produced a common purpose among tribal members that overlay the many differences that existed between tribal members. Throughout the middle decades of the nineteenth century, the tribes flourished in Indian Territory. By the end of the nineteenth century, however, the erosion of tribal independence limited the tribes' ability to control disagreements and rivalries within their populations. And as the United States closed in on the territory, the once healthy competition among diverse interests turned malignant.[3]

* * *

The Indian Territory which Porter fondly recalled was a place the approximate size of South Carolina. The territory's rich and expansive lands presented a variety of economic opportunities. The region's western reaches, including large portions of the Creek, Seminole, Choctaw, and Chickasaw nations, as well as the Cherokees' western precincts, boasted vast grasslands. Magnificent stretches of bluestem provided prime grazing land and became a much-coveted waystation for cattle ranchers heading north to Kansas and the railroads. To the east and in the south, several major rivers provided a well-watered geography suited to a number of agricultural crops, including corn, wheat, oats, barley, and potatoes. There were, in addition, ample stretches of timberland. The territory also included substantial ranges of hills and mountains, where the soil was thin and rocky but where mineral wealth, including oil, lay untapped below. In the southeasterly corner of the region, where the Chickasaws and Choctaws resided hard against the borders of Arkansas and Texas, nature's bounty included extensive coal fields as well as rich soils that later lured cotton growers.[4]

Upon this great expanse there lived by 1850 roughly 45,000 tribal members, a number that remained relatively stable over the next half century. In 1890, there were 22,000 Cherokees, 11,000 Choctaws, 10,000 Creeks, 5,200 Chickasaws, and 1,800 Seminoles, for a total of 50,000.[5]

The cultural dispositions and economic practices of Five Tribes members reflected differences among tribal members that had been evolving for decades. In broad terms, tribal members were divided along class lines into two groups: "small-scale subsistence farmers" and "large landholders."[6] Some scholars prefer different language in their classification of tribal members, suggesting that the term "conservative" or "traditional" replace "small farmer," and "progressive" stand in for "large landholder." These labels, whatever their distinctions, reflect tribal members' attitudes toward the economic development of the territory as well as their relations with white society. In addition, observers of the Five Tribes, as well as of other native groups, have at times used terminology relating to ancestry in describing the differences existing among tribal members. Conservatives, according to these calculations, are the rough equivalent of "full bloods" while progressives are "mixed bloods."[7]

In broad-brush historical treatments of the fundamental fault-lines lying beneath tribal organization, small farmers played the role of the traditional Indian suspicious of Americans and American society. According to one student of the Five Tribes, these tribal members "remained culturally and socially insulated and isolated from the increasingly commercialized economy of Indian Territory" of the late nineteenth century.[8] Small farmers existed outside of the money economy and "continued to exhibit traditionalistic labor ethics—that is, if they received treaty payments, they preferred to engage in the round of community social and ceremonial activities and put off work until it was required to supply their customary necessities of life."[9] These tribal members preferred minimal relations with white society and pursued lives that revolved around communitarian values. Although they adopted technological innovations developed outside the territory, not to mention Christianity and a measure of formal schooling, many small landholders preserved native language skills and traditional religious ceremonies.

As late as 1875, small subsistence farmers made up at least one-half of the tribal populations. They were content with limited agricultural pursuits and grew corn and grains and other necessities for their own use. Many raised livestock. As Union Agent S.W. Marston reported in 1878, "to state the names of all the owners of domestic animals on this reservation would be to give the names of all the men and women and children, for they all have stock."[10] Many of the animals ran free in heavily wooded areas and on unfenced pasturage.

The remaining tribal populations consisted of individuals who operated middling-sized farms, worked at trades and crafts, and pursued otherwise diverse livelihoods. By the 1870s, these tribal members were wary of American forays in Indian Territory, but they were not as adverse in their reaction to the creeping westward movement of white society. By contrast, large landholders, who constituted an elite class within the tribes, were familiar with non-Indian social values and embraced the emerging market economy. While still in the Southeast prior to removal, the large landholders owned slaves of African descent and exhibited other forms of southern culture. In Indian Territory, they nourished a rich institutional life. Among the endeavors that reflected the influence of white society upon their own ambitions, large landholders encouraged the finance and construction of schools operated by the tribes that by the late nineteenth century were reputed to be among the better education centers in the Southwest. Education was provided for boys and girls and the best students often proceeded on to college. At Tahlequah, the Cherokee capital, Bacone College boasted a curriculum that included Latin and the literature of Western civilization. Progressive tribal members received instruction in English and, while vocational skills were taught, there was an emphasis on academic achievement. In the 1870s and 1880s, the educated children of large landholders formed a budding professional class; they were doctors, lawyers, and educators.[11]

Large landholders took advantage of the many opportunities offered by tribal control of valuable assets. In Indian Territory, particularly after the Civil War eliminated slavery, progressive tribal members leased land to non-Indians. Leasing pasturage to Texas cattlemen became an attractive business in the 1860s and 1870s. For relatively little effort and expense, tribal members reaped significant profits, an arrangement that benefited both them and the ranchers who needed grasslands for their herds. Tribesmen also prospered by letting out smaller acreages—fields or shares—to non-Indian tenants. As landlords, they netted handsome profits from the toil of immigrants moving west.[12]

A number of tribal members consolidated control over huge expanses of land and collected impressive leasing fees. Liberal "use" rights within the tribes placed few limits on either the amount of land a person might utilize or the purpose to which it was put. In the Creek nation, for example, each head of family was allowed to fence pastures of up to one square mile without incurring a fee. Those wishing larger tracts could do so at a cost of five cents per acre. As a result, a few powerful families soon controlled great stretches of the Creek nation. Some acquired holdings of thirty thousand acres or more. By 1896, sixty-one "individual citizens or companies of citizens" held more than a million acres of land, or "approximately one-third of the entire area of the Creek Nation."[13] A similar development occurred among the Cherokees where a few families exploited tribal laws to control roughly 175,000 acres by the 1890s.[14]

Even more rewarding prospects beckoned as corporations began seeking business opportunities in Indian Territory. Railroads, for example, competed vigorously for rights-of-way through tribal lands. Enterprising Indians saw that the rich timber lands of the eastern reaches of the territory held out potential profits for sale as railroad ties, and they also understood that the workers laying the rails would require goods and services during their stay in the Indian nations. Companies in search of coal, and later oil, also offered increasingly attractive financial prospects to tribesmen. So long as control over land was not threatened, tribal members welcomed economic growth.[15]

Although they displayed attitudes that spanned the range of sophisticated cosmopolitanism to fierce localism, members of the tribes relied for their well being upon what Porter called the Indians' "Arcadian government." Whether small farmers or large landholders, they possessed a vital commonality that underlay the cultural diversity and economic stratification that existed among them. Even though individual tribal members responded in widely varying ways to the increased presence of non-Indians in their midst and to the beginnings of a sea change in the nature of the American economy, they depended upon their independent, sovereign nations for their wellbeing. Tribal control allowed tribal members to live as they chose and to have access to important resources. Self rule and common landownership, the "integrating influences" of tribal life, bound tribal members together, however different their world views.[16] The shared institutions of nationhood benefited all tribal members. As a result, they sought to preserve their position from outside influence and labored to hold internal disagreements in check lest they invite tribal collapse and their undoing.

The constitutional governments that ruled the Five Tribes, beginning with the Cherokees in 1827 and stretching to 1867 when the Creeks adopted a constitution, promoted the common interests of tribal members. Citizens were vested with property rights that ensured their access to resources and political rights that allowed them to vote in tribal elections. Tribal constitutions, with some variance among them, called for a principal chief or governor, a tribal council, and a representative body or bodies.[17] Tribunals, courts, and judges imposed rule of law, and lighthorsemen provided enforcement. Clearly modeled after the governments of white society, tribal governing institutions separated religious from civil authority. Yet they also preserved important cultural aspects of formerly decentralized social organization. Historically, tribal members had lived in towns held together by complex systems of family and clan reciprocities. These towns formed the basic political units of everyday life, wherein members shared decision making power and rights to utilize land. After tribal government was centralized in response to European and American encroachment, representatives were generally selected by tribal towns, and local leaders retained a measure of power. The tribal constitutional governments blended tradition and innovation thereby rendering them subject to the customs of the tribal polities while granting them legitimacy to carry out treaty negotiations and other tribe-wide matters.[18]

Although bound together through government, tribal members nonetheless vied tenaciously for control of their institutions. Spurred by their diverse economic and cultural concerns, competing interests sought to influence tribal affairs. As a result, tribal government functioned to impose limits on conflict by channeling divergent interests toward compromise solutions. This had not always been the case, and there were several spectacular examples in which tribal divisions propelled the tribes into chaos and even physical conflict. Tribal institutions were susceptible, unsurprisingly, to events beyond their control. During the nineteenth century alone, cases of extreme foreign threat produced upheaval and a consequent failure of government.

During the removal crisis in the Southeast in the 1820s and 1830s, for example, American demands for land ruptured tribal relations and initiated a painful period of disorganization. When legal efforts to retain ancestral tribal homelands in several southern states failed, tribal factions engaged in bitter, even violent, campaigns against one another. For years after their forcible exile to Indian Territory this deep-seated factionalism persisted and was only barely suppressed by the necessity of establishing the tribal nations anew in an unknown land.[19] The fragility of tribal politics was exposed thirty years after removal during the Civil War, when tribal governments faced intense stress. As the Union faltered, both North and South pressured the tribes to declare allegiance and render support. The subsequent loss of independence produced fierce trouble within the tribes, and Indians joined either side, battling one another in vicious guerrilla fighting.[20]

The consequences of the Civil War in Indian Territory, including the signing of new treaties with the United States, had lasting effect on tribal governance. Unrest among the tribes stemmed from disagreements about leadership and from the role and type of government best suited to dealing with whites. But during the difficulties of the past-war period, government remained the common arena in which tribal members pursued their diverse interests. Even as tribal differences became more firmly outlined through the development of political parties, tribal government itself was not undermined. In fact, it was likely strengthened. Under banners such as "National," "Progressive," and "Loyal," political parties competed for adherents throughout the tribes. Small farmers and large landholders supported those organizations that promoted candidates friendly to their concerns. In some respects, by drawing members together in common cause, parties fostered a sense of nationhood. In turn, tribal politics exhibited a fundamental commitment among divergent tribal members to the "nation" and the preservation of its sovereignty.[21] Although older town structures continued to promote fierce support among Indians, the tumultuous era of the 1870s to 1890s moved the decades-long centralization of tribal governance ahead.

Government benefited all tribal members even as they sought different objectives. Small farmers saw that a strong tribal nation was the best chance to protect their way of life and their access to resources. Large landholders enjoyed the liberal "use" rights that were available under tribal law and the fine facilities for education that they had helped build. Both sides understood that to live under any other authority would drastically inhibit their interests. This shared understanding was evident in tribal responses to a number of problems that arose during the late nineteenth century as non-Indians pushed forcefully into Indian Territory.

Ever since 1866, when the tribes signed a new round of treaties with the United States as a consequence of their alliance with the Confederacy, tribal members had disagreed about allowing the railroads into the territory. Although Congress demanded two rights-of-way from the tribes, one running north-south and the other east-west, each of the tribes contained unbending opposition to the actual building of the roads. Small farmers feared that the railroads would open the territory to white settlement, that railroad workers would take up land when construction ceased, and that those who followed once the rails were in place would remain as squatters or tenants. Small farmers also worried that the railroads would initiate the privatization of land. As early as the 1860s, the United States issued warnings of the eventual re-organization of Indian Territory and the dismantling of the tribal estate for the purposes of creating individual homesteads. On the principle that any relaxation of defense might swing wide the door to white domination, small farmers moved to halt the laying of railroad tracks.[22]

Among the Cherokees, tribal factions carried their seemingly contrary views of the railroads to the ballot box and into the tribal council. The more conservative elements of the tribe joined together under the banner of the "National" party and campaigned to halt railroad construction throughout the 1860s and 1870s. They were pitted against members of the "Southern" party, which consisted of larger landholders who "favored the building of railroads for economic reasons, as a way of integrating Indian Territory into the American economy and creating business opportunities."[23] Large landholders also investigated the possibility of acquiring ownership of any railroad built through the Cherokee nation, a plan ultimately forbidden by the federal government. As the two parties spoiled for a fight, the railroad issue loomed large in tribal elections throughout the 1870s.

But contrary to expectations, the railroad issue actually united the two Cherokee factions. It became increasingly clear that while tribal members differed as to how economic development would effect the tribe, the most important matter was to retain control of common resources. And in this, the system of tribal governance tended to foster compromise rather than a showdown in which the winner took all. When Dennis Bushyhead, a Baptist minister who had a large cattle ranch, was elected principal chief in 1879 and again in 1883 on the strength of the conservative vote, he downplayed resistance to opportunities for economic development that accompanied the railroad while simultaneously trying to limit the influence of the railroad companies in Cherokee country.[24] He recognized both the importance and the dangers of the railroad; above all he sought to preserve tribal authority over any and all foreign agents.

Similar patterns emerged in the other tribes. Confronted with Congressional sanction of a railroad through the Choctaw nation, one group of conservative tribal members initially resisted. They tried to tear up rails and filed suit against the railroad company to prevent further construction. But conservative Choctaws soon reconciled themselves to the reality of the railroad and turned to their "government institutions to protect their nationalistic and cultural interests."[25] Among their efforts was to lobby for legislation to tax the railroad companies and to regulate future construction, plans that were supported by other tribal members as well.

Railroad disputes illustrated the signal importance of tribal government in mediating competing interests. While different groups attempted to gain control of tribal institutions, they also fought for the broader preservation of sovereign authority. Large landholders welcomed the railroad for the economic opportunity it promised, but they expected that the tribe as a whole would benefit most. They considered pursuit of their own ambitions to be consistent with the security of tribal rule. Similarly, small farmers in the years after the railroads were built anticipated extending tribal influence over outsiders.

Another area of tribal debate that revealed the central concern of preserving sovereign authority was white immigration to Indian Territory. Drawn westward in search of opportunity, settlers, many of them from the South, excited problems within the tribes. For Indian landlords, white settlers provided a source of cheap labor. Hoping to obtain land of their own, many of those arriving in the territory became tenants and sharecroppers.[26] Unlike other parts of the country, Indian Territory exhibited an Indian elite and a white underclass. Travelers passing through the region noted, often with astonishment, that poor, illiterate whites toiled as tenants on the holdings of wealthy, educated Indians. Tribal landlords, in turn, viewed themselves as belonging to a better, more refined class than the illiterate refugees streaming westward. As the scholar Daniel Littlefield, Jr., has written, "The economically prosperous and educated classes of Indian citizens looked at . . . whites with disdain, forming conclusions about their character from their shiftless nature, their dishonesty, their poverty, and their willingness to let their children grow up in Indian country without an education."[27]

For other tribal members, however, the rapidly growing number of whites threatened to bring ruin. Many recalled the Removal Era when non-Indian settlers thickly populated tribal lands, eventually proving an uncontrollable force that brought about dispossession. Once again it appeared that whites were working busily to unseat the rightful owners of the land, and many small farmers among the tribes fretted that the employment of these intruders as laborers and tenants was "an invitation to American squatters on the common domain, and therefore a threat to national political autonomy, because too many Americans settling on the common domain would surely make it difficult to keep self-government."[28]

Small farmers rallied political support to preserve Indian dominance over whites through institutional controls. They proposed firm limits on whites entering the region and enacted legislation designed to discomfort their stay. Tribal members went to significant lengths to establish a system to ensure that "intruders," as non-Indians in the territory were called, had few opportunities in the territory. Residency taxes were levied and intruders were forbidden to own land. Tribal councils required stiff license fees for those who wished to operate businesses in the territory and barred the children of settlers from attending tribal schools. They also took steps to limit the influence of those men who had taken Indian wives, often for the express purpose of gaining citizenship in the tribal nations and thus gaining access to resources otherwise denied them.[29]

As early as the 1870s, when the number of white settlers in Indian Territory was still quite small, segments of the tribal populations attempted to stiffen the criteria for granting citizenship. Non-Indians, unless they were married to tribal members, possessed no legal standing in the territory and were treated as aliens in a foreign country. The tribes reserved the right to grant whites residence, and one tribe, the Choctaws, even required references attesting to good character before allowing whites to stay, temporarily, in their nation. For those whites wishing to marry into the tribes, regulations soon proliferated as well. Among the Cherokees, "[w]hite and black U.S. citizens had to get 'permits' or marry a Cherokee citizen in order to be allowed to stay in the Cherokee Nation."[30]

As immigration to the West surged in the 1880s, legislation defining citizenship and its accompanying rights reached critical levels. Conservative tribal members wanted strict boundaries between themselves and whites, boundaries that had become disturbingly blurred as the number of intermarried whites grew. Many of these whites sided politically with large landholders and progressive-minded tribal members. At the urging of conservative tribal elements, the tribes passed laws limiting the rights of these non-Indian citizens.[31] After 1877, whites who intermarried with Cherokee citizens no longer automatically acquired property rights. The Creeks passed a law in 1882 "that withdrew the citizenship rights of intermarried Americans."[32] The Choctaws acted with similar directness. According to one historian, the Choctaws "required the applicant [for citizenship] to furnish a certificate of good moral character signed by ten Choctaw citizens, to pay a license fee of one hundred dollars, and to renounce the protection of the laws and courts of the United States."[33]

But despite the laws written and passed by the tribes, by 1890 whites outnumbered Indians in the territory two to one. This development particularly struck panic into the conservative segments of the tribes. They feared white domination and the loss of their autonomy. Unlike progressives, conservatives had little confidence that they could easily navigate white society. A substantial white presence meant pronounced change in their lives and their world. Moreover, conservatives were aware of a shift in political power that was ongoing in the tribes. Large landholders were joining more frequently with intermarried whites to form political alliances. The doors preventing these intermarrieds from gaining a foothold in tribal affairs had swung closed too late. As a result, the influence of progressive tribal members was disproportionate with their total numbers, and they appeared increasingly powerful.[34] They occupied tribal offices and managed substantial tribal resources.

The positon of large landholders in the tribes was growing complex. While they had participated in passing laws to keep intruders out and to retain a firm hold on collective resources, they had begun to form alliances with intermarrieds as the reality of white society's visibility in the territory grew. They manuevered to capitalize on a fluid situation, to gain power for themselves. Yet this manuevering did not augur surrender or a willingness to abandon their sovereign nations. They, as much as their conservative fellow tribesmen, were loathe to give up autonomy, perhaps more so because they had engineered their ascendancy in tribal affairs. However differently they often saw the matter of increased white immigration, large landholders understood that the Indian nations needed to maintain authority lest their resources be exploited and their lives re-arranged. Indeed, tribal members were sensitive to the fact that their own internal disagreements abetted the forces of white society in imperiling their sovereignty. They had to look no further than the Chickasaw nation for an example of how factional wrangling invited destruction.

The Chickasaws, somewhat slower to confront the menace of intermarried citizens, suffered for their "reckless generosity." By 1890, intermarried whites "monopolized the best agricultural lands in the Nation."[35] To correct this problem, the Chickasaws belatedly passed a law that denied property and political rights to intermarried citizens. But they discovered that their prior laxity had allowed whites to organize into a powerful force. Whites held meetings at which they defied the tribe's attempts to limit them. They then threatened the Chickasaws, vowing to "'exterminate every member of this council from the Chief down.'"[36] By 1900, the Chickasaw Nation counted within its borders 150,000 noncitizen whites, 5,000 blacks, and only 3,100 Chickasaws.[37]

Although tribal members worked to curb destructive factionalism, in a few instances compromise failed and conflict boiled over. Among the Creeks, a dispute over election returns in the 1880s, which grew out of a deeper antagonisms regarding the legitimacy of the constitutional government, quickly escalated into an armed standoff between factions at odds about the effects of white intrusion in the territory. This standoff, subsequently called the Green Peach War, deteriorated into a series of skirmishes and seriously challenged the stability of the Creek government. Only a quick response by United States military forces prevented bloodshed.[38] Stemming from similar circumstances, the Locke-Jones War erupted among the Choctaws in 1892 and led to vicious factional violence. Fearing that a national campaign to allot Indian lands would soon arrive in Indian Territory, a number of Choctaws argued that the pace of incorporation into American society was proceeding too quickly. Citing treaties from the 1830s and 1860s, they maintained that only the firmest stance against the efforts of the federal government to implement plans to privatize landholding in the territory should be considered. Consequently, they went to the polls to support a platform of disassociation with whites. After losing to a candidate representing the "Progressive party," an organization that favored development of the territory, an angry collection of Choctaw conservatives took up arms. In the following year, they "plotted to assassinate a large number of progressive leaders," a plan that resulted in several killings.[39]

These incidents revealed the failure of tribal government to harness factionalism. While divisions within the tribes were clear and longstanding, the violent episodes of the 1880s and 1890s indicated the fragility of tribal structures under the weight of external pressures. Shaken by the incursions of non-Indian settlers and other agents of American society, tribal factions spun away from one another, undermining the common ground they had built their lives and institutions upon. As conservatives grew desperate, fearing a permanent loss of influence in tribal institutions, and thus over developments in the region, they concluded that there existed no other recourse but forcible action. They ventured outside the conventional bounds of tribal decision making, weakening the structure of tribal governance.

Yet, as one scholar has noted, internal confrontations such as the Green Peach and Jones-Locke Wars also indicated, paradoxically, support for tribal institutions. Although the risk of bloodshed was great, these conflicts were born of a commitment to, and not a protest against, tribal government. Even the tensest of armed confrontations were not aimed at toppling the established governments. They were, instead, attempts to gain power *within* tribal institutions. They were acts that demonstrated the deep investment of various tribal factions in their nations; tribal institutions were worth the risk of physical harm.[40] Accordingly, while the Creek and Choctaw troubles hinted at increasingly turbulent internal conflict and fragmentation, there existed an underlying unity among tribal members. In the 1890s, even as the embers of the Jones-Locke War still glowed, this unity was displayed when tribal divisions gave way once more to a dedicated defense of tribal resources and autonomy.

But the Green Peach and Jones-Locke Wars also signaled the dangers of the tribes' longstanding internal divisions. The conservatives' resort to arms breached the adversarial yet agreed-upon system of tribal politics that had evolved during the course of the nineteenth century and demonstrated, even presaged, the potential for chaos in the event the tribes were deprived of their autonomy. Indeed, as the new century beckoned, the tribes confronted another challenge that, even as it revealed a stubborn dedication to tribal unity, exposed fault lines that could not be easily erased or covered up.

Having instituted reforms among other Indian groups previously, the federal government dispatched a commission to Indian Territory in 1893 to implement a number of mandated changes for the Five Tribes. Under the leadership of Henry Dawes, the retired Massachusetts senator and principal architect of the federal government's policy to allot tribal land, this commission was sent to overturn the tribes' prior exemption from policies designed to end communal landownership and eliminate tribal structures in favor of preparing tribal members for individual American citizenship.[41] Although no longer active as a lawmaker, Dawes remained a well-connected reformer concerned with the intractable "Indian problem." He came west as chairman of the three-man Dawes Commission to secure agreements with the Five Tribes that would supersede older treaties which guaranteed tribal autonomy and exclusive title to the territory.[42]

Dawes expected swift capitulation. After all, he considered allotment and the subsequent Americanization of tribal members to be a humanitarian policy designed for the benefit of Indians. He further anticipated that the tribes understood the inevitability of American expansion. In all, he presumed that the tribes would be eager to accept his terms as their best hope. But Dawes was mistaken. His hope for expeditious negotiations was dispelled in the atmosphere of suspicion and hostility with which tribal leaders greeted the former senator and his fellow commissioners. For two years the commission strove to obtain new agreements with the tribes—even to bring them to the table for heartfelt discussion. The tribes at every turn "evaded, delayed, and resisted negotiations." Dawes, incensed, grew so intemperate "that he recommended that Congress simply overthrow existing treaties and terminate the tribal governments."[43]

While the ostensible rationale behind the commission's reform policies included lofty ideals of Indian improvement, Five Tribes members viewed Dawes' stipulations as clear threats to their institutions and resources, and thus to all the benefits they had derived in the past. Behind the euphemisms of progress and advancement, tribal members sensed more pragmatic and pressing demands. After Oklahoma Territory—which lay to the West of Indian Territory and was comprised of those tribal lands handed over to the federal government in the wake of the Civil War—was opened to white settlers in the spectacular first land rushes of 1889, popular resentment toward Indian Territory had increased.[44] There was escalating demand that the two territories merge to create a new state.[45] The tribes knew that their very existence in Indian Territory posed the most significant obstacle to this plan. The Dawes Commission, it was evident, had come to tear down what the tribes had built.

The stalemate between the tribes and the commission ended when the federal government provided Dawes with crucial leverage. In 1895, surveys of tribal land were authorized, followed the next year by legislation ordering the compilation of tribal census rolls, a precursor to the division and disbursement of individual lands. Although they continued to drag their heels in confrontations with the commission, tribal leaders by 1898 were compelled to conclude matters and to accept Dawes' terms. In that pivotal year the Seminoles signed an agreement with Dawes, and the Curtis Act, which gave Congress power to end tribal land tenure without tribal consent, was passed. In short order, the remainder of the tribes agreed to compacts under which they relinquished common ownership of land as well as their sovereignty. Their united defense of tribal powers and resources was shattered. Under the new agreements, tribal members would be granted individual allotments and their governments were slated to cease their functions in either 1905 or 1906.[46] In time, the Five Tribes would be absorbed by the United States and their former nations incorporated into American society.

But the Dawes Agreements went beyond the opening of the territory for settlement. They undid the complex connections that held tribal members together, driving a wedge between tribal factions and separating Indians from one another. The unity displayed during 1893 and 1894, when the tribes dodged Dawes and his fellow commissioners, stalling them and evading their entreaties, crumbled. In its place, turmoil arose. The long evolving competition between tribal interests, competition that had occasionally flared into hostility and violence, was now unleashed. What unity existed previously was breached by strident accusations of betrayal. Small farmers and more conservative-minded tribal members felt that the Dawes Agreements constituted tribal suicide and those who signed them traitors of the ugliest persuasion. They argued for the sanctity of tribal institutions and common landownersip. Large landholders, in contrast, assumed by the late 1890s that allotment was inevitable and that to resist was foolish. Given their own abilities to prosper, they expected that they would continue to do so when the region was incorporated by the United States.

* * *

Pleasant Porter understood the fragility of the bonds that existed among tribal members. In 1906, he appraised the destructive force that the Dawes Agreements visited upon tribal relations:

> There is that sense of right and wrong which will bind men together and preserve the peace and maintain virtue and provide for offense without. That is the institution out of which a nation grows. Each of these groups [the Five Tribes] must have had that; but you rub that out, you transplant them into what they have no knowledge of; . . . *there is no life in the people that have lost their institutions* [emphasis mine].[47]

THREE
"Real" Indians

In 1906 Jacob B. Jackson, a Choctaw, argued for the right of his tribesmen to move from Indian Territory and find a new home for themselves, perhaps in Mexico. Appealing to a Senate committee, Jackson made clear that conditions in the territory demanded drastic action. He was especially concerned for the racial integrity of the tribe, which he considered to be imperiled by the Dawes reforms and the subsequent flood of white settlement. "Surely a race of people," Jackson told the visiting senators, "desiring to preserve the integrity of that race, who love it by reason of its traditions and their common ancestors and blood, who are proud of the fact that they belong to it may be permitted to protect themselves, if in no other way by emigration." Jackson went on to compare the desire of tribal members to emigrate with that of an English colonist's voyage to the New World, when he "went across the great ocean and sought new homes in order to avoid things which to him were distasteful and wrong."[1]

In explaining further his reasons for supporting emigration, a scheme discussed by a number of groups within the Five Tribes during the first decades of the twentieth century, Jackson paid careful attention to the changes that not only were re-making the landscape and thus forcing tribal members to consider a second westward removal, but to the changes that threatened to alter the Indians' own bodies. He alluded to the importance of ancestry in discussing Indian matters. "If the Choctaw and Chickasaw people as a whole were willing to lose their racial status," he argued, "to become by a slow process of blood mixture, and through changed conditions, white men in fact, . . . we do not oppose the carrying out of their desires." But "as an Indian, he had certain rights, among which is a right to exist as a race, and that in the protection of that right, it is our belief that we are fulfilling the purpose of the Divine Creator of mankind."[2]

Jackson was not theorizing abstractly. He understood that full blood Indians were, if not becoming fewer with each year, comprising a declining proportion of tribal membership. As of 1875, they represented roughly one half of the tribal populations, or 25,000 people. By the time the Dawes Rolls were closed in 1907, the number of full bloods—which legally meant tribal members who could prove a blood quantum of more than three-quarters Indian, as well as tribal, parentage—was listed as 26,774, or thirty five percent.[3] For Jackson and others, the case for emigration was an extension of efforts to defend full blood survival. The "slow process of blood mixture" appeared inevitable if the tribes remained in the territory. Even for those who did not entertain removing to Mexico, the threat of racial amalgamation pushed them to consider ways to protect themselves, either through resistance to white society or through new organizations dedicated to preserving tribal structures and the remaining "real" Indians.

In championing blood, conservative tribal members aggravated the divisions already present in the tribes. By conflating Indianness with behavior, maintaining that "real" Indians were those with a full complement of Indian blood, tribal members endowed questions of legitimacy with added significance: tribal members who assented to the destruction of tribal institutions were open to suspicions about the nature of their claim to an Indian identity. How could they be "true" Chickasaws or Seminoles if they helped end the Chickasaw or Seminole nations? Progressives who supported the apportionment of collectively-owned lands and the dismantling of sovereign governments were accused of forsaking their responsibilities; and as mixed bloods they were illegitimate spokesmen. In contrast, "separatist" organizations, comprised largely and self-consciously of full bloods, emerged during the late nineteenth and early twentieth centuries to wrest affairs from those who consented to the Dawes Agreements.

* * *

Conservative decline, already evident by the 1880s, was exacerbated by the Dawes reforms. Under federal policy, tribal institutions were to be disbanded as a precursor of statehood. The Dawes Commission dispatched enrollment parties and surveyors to carry out allotment in each of the tribes. Tribal members were assigned homestead lots and were instructed to build homes and to farm. With their "surplus" lands, they were encouraged to lease to non-Indians or even to sell. But their access to commonly-held timber lots, grazing pasturage, and other benefits evaporated as allotment certificates were distributed. Tribal members were instructed to acclimate themselves to conditions under which they would become individual citizens. For conservative tribal members, staunch defenders of tribal structures, the loss of self-rule threatened a way of life. Customs, values, and practices guarded by tribal government were to be exposed to a less than merciful white society.

The Dawes Agreements contained the seeds of disaster. Conservative tribal members relied heavily on institutions in which they had a say and which were predicated upon communal values. Most troubling, it appeared that their tribal opponents had carried the day in the negotiations with the Dawes Commission. During their conflicts of the late nineteenth century, conservatives and progressives managed to provide a balance for one another, to ensure that neither group might gain dominance. But the affinity that many progressives had for white society and the ease with which some progressives were already moving within white society plainly suggested that the Dawes reforms favored them. Progressives exploited their position to the detriment of other tribal members. Conservatives, who maintained a distance from whites, feared severe dislocation as a result of the new policies.[4]

Conservatives responded to these developments in various ways, including the formation of "separatist" groups designed to both resist the Dawes reforms and to reaffirm tribal autonomy. In each of the tribes there existed movements among small farmers and other conservative-minded tribal members to re-establish independent government and to take action against those who argued otherwise. Among the Creeks an organization called the Snakes assailed the work of the Dawes commissioners and actually created substitute institutions to take the place of those that were being extinguished. Led by Chitto Harjo, a charismatic Creek also known as Crazy Snake, the group rallied together at the town of Hickory Ground where in 1900 they created a "shadow" government; they held their own elections, appointed their own judges, and carried out their own brand of justice.[5] They posted notices warning non-Indian settlers to leave the area and issued death threats to the Dawes commissioners. Mounted and armed lighthorsemen patrolled the roads in the vicinity of Hickory Ground periodically terrorizing travelers and farmers.[6]

Chitto Harjo regarded white society with contempt, but his most trenchant criticism was reserved for Creeks he deemed complicit in assisting the Dawes Commission. Those who supported federal plans, he maintained, had betrayed the Creeks; they were traitors. As retribution, the Snakes decreed that tribal members discovered with allotment certificates were to be whipped and fined. Those who leased or sold their lands or employed non-Indians to labor in their fields faced similar punishment. Two Creek leaders, Isparecher and Pleasant Porter, both of whom negotiated with the Dawes Commission in 1898, were singled out. Fences on Isparecher's property were torn down and Porter received death threats.[7]

The Snakes sent emissaries to Washington and urged that Congress recognize the inviolability of older treaties. When those efforts achieved scant success, they took matters into their own hands at Hickory Ground. The staunch defiance of the Snakes was of such a quality that even vigilante raids by neighboring whites and repeated arrests by federal marshals could not stamp out Chitto Harjo's followers. The Snakes continued to agitate, and the troubles at Hickory Ground lasted throughout the first decade of the twentieth century. In a 1905 interview with the Creek writer Alex Posey, Chitto Harjo echoed the language of older treaties guaranteeing the tribe its lands indefinitely: "I shall never hold up my right arm and swear that I take my allotment of land in good faith—not while the water flows and grass grows. . . . I notice that the grass is still growing, that the water in the North Canadian [River] is still flowing toward the sea and that the leaves still appear upon the trees."[8]

Chitto Harjo's example spurred members of other tribes to action. Emboldened by the Creeks' efforts, Choctaw sympathizers, also known as Snakes, temporarily ousted the existing Choctaw tribal council in 1901. Two hundred Choctaws in Gaines County announced their seizure of the rightful Choctaw government and forced the principal chief to flee to Arkansas. These rebels then held elections for a chief and council and "exercised governmental functions." Elsewhere, "Seminoles sympathetic to the [Creek] Snake cause attempted to enact a code of laws similar to the Snake statutes."[9]

Among the Cherokees, an organization known as the Keetoowah Society maintained that older treaties, dating to the 1820s and 1830s, had been flagrantly ignored by accommodationist tribal leaders. In 1901, the Keetoowahs criticized the Cherokee government for its meekness in negotiations with the Dawes Commission. Meeting at Tahlequah in 1901, the society passed resolutions against the Cherokee National Council. The group was critical of provisions that the council agreed to, especially the division of the tribal domain. The society also faulted the Cherokee government for agreeing to the enrollment of tribal members on official government census lists without consent of the entire tribe.[10] Numerous Keetoowahs evaded the enumerators assembling the tribal rolls, known as the Dawes Rolls, and they refused their allotment certificates. Others, who allowed themselves to be enrolled and allotted, asserted in various resolutions that acceptance of their assigned homestead parcels did not signal their approval of the abrogation of former treaties.[11] They continued to protest and to support organizations such as the Keetoowahs in efforts to reverse the Dawes reforms in Congress.

Like the Creek Snakes, the Keetoowahs attempted to negotiate directly with the federal government, thereby sidestepping the Cherokee government. In 1901, reminding the Dawes commissioners of the "devotion and faithful service of the Keetoowahs during the" Civil War, the Society, which claimed upwards of 5,000 members, explained that "we do not receive [the Dawes Agreements] with joy and gladness, but with the deepest sadness instead . . . and surely we will not be blamed if we decline to take an active part in this annihilation."[12] Keetoowah deputies were sent to Washington, DC., and a series of memorials were forwarded to federal officials. With the Cherokee government scheduled to expire in 1906, the Keetoowahs independently sponsored an election in 1905 for a new principal chief. Unhappy with the current elected leader, W. C. Rogers, a progressive-oriented Cherokee, the Keetoowahs elected an oppositional slate to the national council and Frank Boudinot, an attorney who represented the group, as chief.

Although the election was sanctioned neither by the federal government nor Rogers, "the action," according to one writer, "demonstrated the independence of the Keetoowahs and the belief that they represented the will of the true Cherokees."[13]

Tribal separatists maintained that they were different, not only from whites and white values but from other members of their tribe. And this particularity was both exposed and intensified by the Dawes reforms. For the Keetoowahs, allotment promised not a better life, as government officials envisioned, but the loss of precious resources. As people engaged in working the land and participating in the cooperative ethos of tight-knit community life, they considered common landownership vital to their very existence as Cherokees. The distribution of acreage would rip apart their communities and consign Keetoowah values and beliefs to extinction. Kin and clan relations would be undermined by allotment and subsequent white settlement in the Cherokees' midst.[14]

Unlike the tribe as a whole, which was divided along many lines of class and culture, the Keetoowahs viewed themselves as unified by their culture, language, and traditions, not to mention their political outlook. Moreover, in a context of territorial development and upheaval, they shared an economic "class"; they lived in roughly equivalent manner. Keetoowahs were overwhelmingly small farmers who toiled for subsistence. They lived in spare cabins, wore rough clothing, ran livestock, and took advantage of nearby forests for wood and hunting. By the turn of the century, however, as the market and white-dominated political institutions advanced into the territory, this economic strategy was rendered increasingly unfeasible. Keetowah assets were chipped away, and they faced an increasing need to obtain money in order to purchase necessities. Keetoowahs were uniformly frightened by the threats of local and state authorities to impose taxes on Indian lands, a development that would similarly require the cash-poor tribal members to confront a foreign authority with inadequate knowledge and resources. Although the federal government, through the Indian bureau, fought an occasionally successful rearguard action against such taxation, popular attitudes among whites in the region denounced special treatment of Indians.[15]

Ruthless victimization, in addition to other shared circumstances and traits, provided fertile commonality among conservative tribal members like the Keetoowahs. Beyond the grinding forces of economic change sweeping westward, tribal members encountered a rising tide of criminality. Legions of white speculators selected from a veritable grab-bag of swindles and confidence games designed to rob Indians of their allotments. One favored method was the arrangement of guardianship for Indian wards, both minors and other legally defined "incompetents." Guardians used their complete control over wards to steal money and land. Conservative tribal members placed under such "protective" care intimately understood their powerlessness. They knew that few effective protections existed to help them preserve their lands and the meager livelihoods derived from them.[16]

The Keetoowahs understood that the conditions they endured were not shared by all Cherokees and thus their view of tribal affairs were unlike those of other tribal members. In an appeal from 1901, the Keetoowahs requested that the tribal government appoint additional delegates to convey their special concerns to Congress:

> We earnestly solicit your consideration of the fact, well known to all of us, that the end of the Cherokee Nation and final division of property close at hand strikes closer and deeper into the hearts and lives of the Kee-too-whah . . . than to any other people, or class of people, on earth.[17]

In their reckoning, the Keetoowahs represented the more legitimate Cherokees, and their legitimacy was proven by their vulnerability to external forces and by their unusual, defiant devotion to the tribe, its autonomous government, and the principle of common landownership.

Conservative tribal members looked eagerly for others who shared their fears, their poor material existence, and their impotence in the face of white society. Often, they were willing to cross tribal lines to form connections with others. They appreciated the similarities that existed between themselves and conservatives in different tribes, especially when contrasting their values and beliefs with those of white society and those Indians who acted like whites.

Some groups, like the Creek Snakes, actively tried to influence other groups. Not content to lead by example, the group interceded in the affairs of neighboring tribes. Chitto Harjo invited representatives to Hickory Ground in order to spread the gospel of resistance, and he sent supporters to other tribes to promote the anti-allotment cause. "Snake emissaries were sent out to recruit members from other tribes," one historian notes, "and dissident Choctaws, Cherokees, and Seminoles joined the Creek Snakes."[18] The Creeks "welcomed delegates from the 'disaffected elements' of all Five Tribes."[19] Among the Cherokees, members of the Keetoowah Society offered support to Chitto Harjo, and Keetoowahs were frequent visitors to Hickory Ground and other locales where reaction to allotment was festering.[20]

The willingness of conservatives to form links between people of different tribes but similar political positions and material circumstances reached a more significant level with the creation of inter-tribal organizations. Perhaps the best expression of this impulse was the formation of the Four Mothers Society, or Four Mothers Nation. Drawing support from among the conservative populations of the Five Tribes, this loose-knit movement, which at one time claimed upwards of 24,000 members, picked up the threads of various separatist groups and attempted to tie them together.[21] Its original goal was the restoration of former treaties granting the tribes' their land in perpetuity. In 1906 the Four Mothers submitted a petition signed by one hundred and eighty-six members to the government requesting that money be provided for a delegation to travel to Washington where the group planned to find a sponsor in Congress. Despite the fact that former treaties had been superseded by those signed after the Civil War in 1866 and, more recently, by the agreements elicited by the Dawes Commission, the conservatives' strategy was to focus on past wrongs and stress the unique relationship between the federal government and the tribes. They also took issue with the ability of tribal leaders to sign treaties for the rest of the membership.[22]

As separatist groups took root in each of the tribes, and across tribal lines, they imposed increasingly exclusive standards on the concept of "Indianness." These standards addressed differences of economic class and culture; they addressed the behavior of tribal members. Conservatives, people who tilled small plots of land and practiced traditional ceremonies, who cooperated in the annual agricultural work cycle to ensure every family might have enough food and adhered to older religious and social codes of conduct, were deemed "Indian." Those who lived in other ways, who embraced different values, lacked important aspects of Indian identity. This deficiency was manifest in their improper behavior. But separatists were not satisfied with drawing lines between behaviors, with forming identification around matters of class and culture alone. They argued that there in fact resided a deeper, more essential determination of Indianness, one that derived from blood.

Shades of racial difference had long been meaningful among tribal members, particularly given the tribes' centuries-old exposure to other peoples since European and African arrival in the New World. Intermarriage, African slavery, and the granting of tribal membership to various friends and allies had opened the door to a complex racial amalgam.[23] Alongside these developments, the terminology of blood became well established. While still in the Southeast, the Five Tribes had lived in a climate marked by racial hierarchies and designations. The more educated members of tribal society were men and women with ample exposure to the racialist thinking that prevailed among their white acquaintances, thinking that held as a commonplace that certain peoples were more civilized than others.[24] But other tribal members also were aware that differences between the races existed. Scholars have noted that as early as the seventeenth century certain groups of Indians referred to themselves as "red."[25] By the early nineteenth century, observes William McLoughlin in a study of the Cherokees, conservative tribal members claimed racial separation from whites. Tribal origin stories revealed that Cherokees considered that they had evolved apart from whites.[26]

Early efforts to absorb racial thinking, however, were given new urgency by the late nineteenth century. The process of contact, stretching over centuries, and removal, stretching over hundreds of miles, had gone a long way toward eroding tribal strength and identities. Intermingling and intermarriage produced a variegated landscape of widely ranging mixed peoples who displayed a similarly broad number of cultural dispositions. This dizzying array pushed the tribes to incorporate ancestry into discussions of citizenship and marriage. As property "use" rights were a vital route to wealth and power, the tribes used ancestry as a shield to preserve their resources just as they erected other barriers against dispossession by the increasing numbers of whites moving west. To the many laws and regulations designed to separate Indians from those pressing in on them, to articulate standards and criteria for establishing limitations on intruders, tribal members added a novel coda. In the sensitive matters of determining citizenship or granting permission for marriage, evidence of ancestry was required. The tribes demanded proof of parentage and framed their demands in the language of blood. In 1886, to fend off claims by Choctaws who had remained in Mississippi as well as the influx of several hundred people with controversial and contested ties to two Choctaw families, the tribe's legislature questioned the ancestry of the claimants. According to one historian, the Choctaws "decreed that thereafter no claimant would be admitted to citizenship without proof of at least one-eighth Choctaw blood."[27] As for marriage, a white man desiring a Cherokee wife "had to get the signatures of at least ten 'Cherokees by blood.'"[28]

While members of the Five Tribes prior to the 1890s used racial ideas more to separate themselves from whites and blacks than from one another, conceptions of race hardened by the late nineteenth century; racial characteristics appeared less mutable than in the past.[29] Tribal members more frequently applied racial distinctions to one another.[30] As with other aspects of political polarization taking place in the tribes, physical appearance and ancestry contributed to the widening tribal divisions. When Eufaula Harjo, a Creek Snake, suggested that "the Indian people" had no part in approving the Dawes reforms, he knew full well that many tribal members endorsed the plan. His intention was to advance a selective view of who "the Indian people" were. John Kelly, another Snake, added: "The real Indian was not consulted as to allotment of lands," and "if he had been consulted he would have never consented to depart from the customs and traditions of his fathers."[31] The idea that there existed "real" Indians suggested that others were in some way counterfeiters; they lacked a crucial claim to Indianness. That claim increasingly was attributed to ancestry. Chitto Harjo, in explaining his defiance of allotment, stated: "The civilization of the Indian has not changed very materially, he still tills his sofky patch, his *color* remains the same and he attends to his business and has asked no change"[emphasis mine]. Among the Cherokees, the Keetoowah Society often referred to themselves expansively as the true representative of the "full-blood Cherokees."[32] And individuals from throughout the tribes referred repeatedly and authoritatively to racial traits.

Time and again the rhetoric of conservative leaders drew sharp distinctions between full bloods and mixed bloods. In their speeches and writings, conservative tribal members almost always indicated, before anything else, their blood quantum. In a letter to the Dawes Commission in 1901, for example, Keetoowah leaders John McIntosh and Robert Tolen explained that the "Keetoowah Cherokees are nearly all pure blooded Cherokee Indians."[33] Blood cropped up several times throughout their letter and established the legitimate identity of the letter writers and their followers as well as helping to explain their position regarding allotment. Chitto Harjo in 1900 pointed out that he was the representative of "the Muskogee nation of the Ind. Ter. and known as the fullblood Indians of said Muskogee (or Creek) nation."[34] Lesser known individuals also made sure to note their blood, even in the most routine correspondence. Remarking on his opposition to allotment, Osway Porter, a Chickasaw, spoke confidently for a large portion of the population: "I am a full blood and I know what the full bloods want. . . . I know that I have not ever wanted our country to be divided, and I know that none of the fullbloods have ever wanted it divided."[35]

The use of blood by tribal members was not an idle form of introduction. Rather, such a preface in writing or speech signified the centrality of blood as a form of identification. By the first decade of the twentieth century it was understood that ancestry conveyed a wide range of attributes—appearance, education, upbringing, economic status, language abilities, understanding of tribal traditions, and, importantly, political outlook. Blood was a form of shorthand: by relating one's ancestry up front, at the start, the writer or speaker was relating many tangible and intangible pieces of information that were to be noted by the reader or listener; it was an immediate categorizer. When using blood to identify themselves, people committed to an already understood and defined pattern of behavior.

The repeated invocation of ancestry, particularly as a form of lament, underlined conservative tribal members' despair that their position in the territory was, at best, tenuous. Like their land and sovereignty, their very biological distinctiveness seemed to be dwindling under assault by outsiders. Indeed, for years conservative tribal members had decried the rising influence within the tribe of intermarried whites and their mixed blood offspring. By the early twentieth century, the trend appeared overwhelming. Where boundaries between peoples were increasingly blurred by intermarriage and white settlement, just as surely as the boundaries between the independent tribal nations and the United States had been blurred by conquest, conservative tribal members seized on a finer element to differentiate themselves—their history and their character—from others. This element was blood, a concept ironically derived from white society, which was itself locked in an ideological struggle to define and categorize racial hierarchies.[36] If conservatives could establish their claims as "real" Indians, they might be able to reestablish the boundaries that previously divided Indians from whites.

In 1904, a Choctaw Civil War veteran expressed his concern for full blood demise. As reported in the *Muskogee Times Democrat*, Capt. Peter Maytubby speculated on the future of Indians as a distinct people in the region. "Full bloods among the Indians are becoming comparatively few, and I think the day is not far off when there will be no red men no more."[37] Without full bloods, then, there were no Indians. A full blood himself, Mayutbby had worked on the Dawes Commission for four years and knew whereof he spoke. Allotment weakened the fabric of community and impoverished tribal traditions. In large part, this resulted from allotment's tendency to attract white settlers into the region. Maytubby asserted that tribal institutions and communities previously were strong because of the fact that they had been racially unified. To mix tribal members with whites opened the door to racial amalgamation and hence to the loss of cultural structures.

Pleasant Porter, in comments to the Oklahoma state constitutional convention in 1906, was similarly pessimistic about the fate of the full bloods. Although not a separatist like the Creek Snakes, Porter nonetheless expressed pride in his Indian ancestors and came to view the developments of allotment and impending statehood in a negative light.

> The white element and the element that it can control is in the saddle, and in the ordering of things the Indian has neither place nor part. The Indians haven't had time to grow up to that individuality which is necessary to merge them with the American citizen. The change came too soon for them.... There will be a remnant that will survive, but the balance is bound to perish.[38]

This "remnant" did not include those of mixed parentage, but rather the diminishing number of Indians who might retain their unspoiled biological lineage.

Implicit in the use of ancestry was the definition of others as possessing a different, or weaker, ancestry. Accordingly, full bloods viewed mixed bloods less as fellow tribal members than as outsiders who had gained entry behind the gates of the tribal defenses. Events reinforced this view. During the process of allotment, for example, mixed bloods were in a better position to reap individual rewards. One historian has written:

> When the commission to the Five Civilized Tribes opened the Creek land office at Muskogee in 1899, there was a rush to file by those citizens of the nation possessing the least Indian blood. These people secured the cream of the Creek Indian land. Later the full-bloods began slowly to file upon their allotments, but in almost every instance they could find nothing to file upon but second and third grade land. The best lands lying along the streams and adjacent to thriving towns had all been taken up.[39]

In discussing the activities of the Dawes Commission, the Creek Eufaula Harjo accused "half breeds" of cheating the full bloods. "After our country was divided," he recounted in 1906, "they would send the half breeds around—the half breed Indians—they would go out and hunt for the names of the full-blood Indians without their consent, and they would take the names down and go present them before the Dawes Commission, and these half breeds that brought these names before the Dawes Commission would go and take an oath over it." In the end, Eufaula Harjo said, a full blood Creek would receive his allotment certificate in the mail and "not know anything about it." The "half breeds," he maintained, simply lied with impunity, "take an oath over it," thereby sacrificing a fellow tribesman's principles and desires.[40]

The emphasis on race was understood by progressives who interacted with the separatists. In his biography of the noted Creek writer and journalist Alex Posey, the scholar Daniel Littlefield, Jr., observes that Posey, after long experience and repeated interactions with his more conservative tribal fellows, came away with a strong sense of the growing distrust among tribal members. After serving on an enrollment party charged with visiting some of the most remote sections of the Creek nation to obtain census information, Posey recorded "his growing concern for the gulf that existed between him, as a Creek, and them. As he read over his journal notes," Littlefield wrote, "he must have recognized the ample evidence that the Snakes viewed him—this educated young man, dressed in his white shirt and tie, who sought them out in remote cabins—as different from them both racially and socially." Posey received an earful on his various excursions into the countryside from Creeks who talked of the "real" Indians and who referred to "color."[41] Posey himself used the language of race to criticize the Snakes, and he began to refer to them as "has beens" and "pullbacks," and even as "Indians," a term "that took on more racial and social meanings for him as time passed."[42]

Progressives, thus, were viewed with suspicion by conservatives not only because they advocated policies that undermined tribal institutions, but because their support for those policies stemmed from their "unIndian" ancestry. Progressives, in effect, were too white. It could hardly escape notice that many large landholders and other progressive-minded members were the children of mixed marriages between whites and Indians. Children of mixed parentage grew up bilingual and knowledgeable of different sets of cultural values and practices. Whereas this type of cross-cultural insight once served the tribes well, and mixed bloods were crucial in eighteenth and nineteenth century negotiations regarding trade and land cessions, it had, by the twentieth century, become a negative mark. Over time, mixed bloods revealed an inclination to favor white society, and they adapted more easily to the changes accompanying American expansion westward. Mixed bloods were obvious in their success.

In the racial calculus of the late nineteenth and early twentieth centuries, blood assumed determinative powers. First among whites and then among Indians themselves, the behavior of individual Indians had come increasingly to be equated with ancestry. Although the reform efforts of the United States government, promoted by various "friends of the Indian" such as Henry Dawes, suggested that all Indians might be raised up to a greater state of productivity and equality, most whites and government policy makers considered ignorance, laziness, and thriftlessness to be endemic to the Indian race. And the more Indian blood a person had, the more prone he was to the negative aspects of the typical Indian character. Full bloods, according to popular stereotypes, were near-savages. While there were exceptions, and full bloods, if properly nurtured from an early age, might gain a semblance of civilization, they would always be susceptible to lapsing into a primitive state. Full bloods were understood to be mired in poverty and ignorance, or they were specifically targeted as victims precisely because they were full bloods; it was an inherited station. In contrast, mixed bloods were viewed quite differently. Mixed blood achievement, whether a successful professional career or the obtainment of intellectual sophistication, was not infrequently considered the result of breeding.[43] What these stereotypes indicated was that individual tribal members possessed certain unchangeable sets of characteristics; blood established firm boundaries between them.

While white stereotypes based upon notions of blood negatively affected conservative tribal members, they themselves employed blood in constructive, proactive ways. From a different, but no less biologically determinative, perspective, full bloods regarded their Indian blood as a vital aspect of their claim to "real" Indianness. They, unlike mixed bloods, could boast pure bloodlines and hence a strong identity with Indian culture. Their focus on preserving blood constituted an effort to retain a host of attributes of Indianness. In this, former tribal institutions appeared the means by which Indians were able to live distinctly Indian lives. Tribal towns were racial enclaves where Indians harbored the traits that made them Indian. Thus, when these structures came under attack by federal policy and white settlement, blood came to be regarded as imperiled and in need of protection. Indianness was something that required a dedicated defense; it demanded scrupulous identification of the qualities that made an Indian "real" or genuine. To safeguard Indian blood was to promote the continued survival of "real" Indians. As whites flooded into Indian Territory there was a grave fear that Indian blood would be hopelessly diluted.

Tribal members deliberated at length about removing further westward or to Mexico beginning as early as the 1830s and as late as the 1920s. Mexico, in particular, was viewed as a "potential haven from whites," as well as from those tribal members who accepted white society, and some members of the Five Tribes went so far as to scout out lands and inquire as to grants from the Mexican government.[44] Appeals for money to carry out tribal emigration were frequent, and tribal members conducted fund-raising campaigns to finance the purchase of acreage elsewhere. Both Cherokee and Creek conservatives petitioned their respective tribal councils in the years preceding Oklahoma statehood for appropriations to help them emigrate. Seminoles and Choctaws also inquired as to the feasibility of the plans and advised government officials that they were undertaking preparations to move.[45]

Those behind emigration argued for the need to separate themselves from white society and from those tribal members who had become, in John B. Jackson's words, "white men in fact."[46] Contained within these plans was a recognition that the tribes did not depend upon an attachment to the land of eastern Oklahoma as part of their identity. They had moved in earlier times and, if need be, they would do so again. The sense of peoplehood advocated by those who wished to travel further West was one that derived from blood. One scholar notes that the plans hatched by various separatist groups to emigrate were "utopian."[47] While this may be so, those plans strongly indicated the emphasis that conservative tribal members placed on ancestry in determining Indianness.

Conservative tribal members employed blood in the service of separatism. They asserted their distinctiveness on the basis of ancestry and sought to weed out those in their midst who would betray them—or to weed themselves out, as was the case with supporters of emigration. As they formed organizations such as the Snakes, as well as inter-tribal groups such as the tongue-in-cheek named Indian Bureau begun in 1906 by Eufaula Harjo to "represent the interests of all full-blooded Indians in the state," they demonstrated a will to divorce themselves from other tribal members, to exclude those who did not share their own views.[48] Instead of accepting their political rivals as legitimate participants in debate, as they had in the past, conservative groups, sensing that the upheaval in Indian Territory would soon lead to the permanent loss of their way of life, separated themselves from those who disagreed with them. It appeared better to focus on a core of like, and like-thinking, members to save the tribes from ruin.

The effect of these ideas and actions was the creation of limited replicas of former tribal institutions. As their old governments were being swept aside, tribal conservatives strove to replace them with newer, more cohesive substitutes. This was a move away from previous, inclusive forms of tribal governance of the nineteenth century and toward the achievement of separate interests irrespective of those who possessed different beliefs and values. Relying upon blood as a crucial feature of Indianness, conservatives narrowed the chance for accommodation of diverse tribal memberships. As a result, alienation among tribal members accelerated.

* * *

By the turn of the century, not only were the tribes internally divided and externally pressured, but the nature of tribal relations had altered. Instead of drawing together to vie for power within tribal institutions, opposing tribal members now pursued divergent objectives with little reason to reconcile with one another. In turn, tribal members grew increasingly critical of one another, so much so that they began to question fundamental aspects of shared tribal membership. It was no longer clear that all tribal members were equally "Indian." The process of exclusion emphasized like and unlike characteristics, none more divisive than blood. The criteria for Indianness had narrowed, and precious little room remained in which to discuss a range of ways in which it was possible to be Indian.

FOUR
The Importance of Being White

I N HIS INAUGURAL ADDRESS AS OKLAHOMA'S FIRST GOVERNOR IN NOVEMBER 1907 AT Guthrie, the temporary capital, Charles Haskell celebrated the harmonious union of Indians and whites in the creation of the nation's forty-sixth state. Oklahoma, Haskell told the assembled crowd, derived its character from the fact that its founders, despite their diverse cultural heritages, had come together in peace. The state provided a model of race relations, and the governor remarked that the red and white stripes of the American flag might best symbolize the joining of the red and white races. Among the ceremonies that day in Guthrie, a "marriage" was held in which Indian and Oklahoma Territories were united. Standing in as the groom was a white cowboy, the personification of Oklahoma Territory. His bride was an Indian maiden.[1]

The boosterish attitudes on display at Guthrie appealed to many members of the Five Tribes. Having accepted the inevitability of allotment and incorporation of the territory into the United States, progressive tribal members saw the chance for unfettered upward mobility. They considered that statehood might fulfill the aims of Henry Dawes and other policy makers to convert Indians into American citizens. Further, there existed the possibility of racial equality, a matter to which Haskell explicitly referred. Although it might be dismissed as celebratory rhetoric, Haskell's speech revealed an underlying belief that Indians could join with whites in American society.[2] Indeed, the mock marriage ceremony suggested that a deeper intermingling was possible, perhaps even desirable.

Where conservatives saw trouble, progressives saw opportunity. While conservative-minded tribal members rallied together as "real" Indians, progressives adhered to racial categories which ingratiated them with whites. As descendants of mixed marriages, they considered themselves better able to enter white society than full bloods. Mixed bloods also expressed concerns about full blood backwardness as a hindrance to their ambitions. When tribal separatists refused to accept assimilation—created shadow governments and took the law into their own hands—they perpetuated stereotypes of Indians that connoted racial inferiority.

Mixed bloods were sensitive to the power of racial ordering in the region. The early twentieth century produced violent conflicts between whites and blacks in Oklahoma, and segregationist practices were widespread. Progressive-minded tribal members, fearful of association with other dark-skinned racial groups, separated themselves from blacks as well as full-blooded fellow tribesmen.

* * *

While conservative tribal members grew more tribal-minded in the early years of the twentieth century and built new organizations such as the Snakes, progressive Indians sought to compete in American society as equal individuals. They believed that their fortunes were in their own hands. This view was backed by federal and state authorities, which encouraged an individualistic ethos. In contrast, conservatives were perceived as incapable of taking care of themselves.

The Dawes Act of 1887 originally stipulated that Indians could become United States citizens upon receiving their allotments. When applied to the Five Tribes by the late 1890s, this act continued to serve as a way for Indians to achieve their equal place in society. But the federal government appreciated that differences existed among tribal members, and it installed provisions to protect novice landowners by placing allotments under a twenty-five year trust status, during which time they could not be sold or taxed.[3] As statehood neared, the federal government implemented additional protections for certain tribal members while lifting trust restrictions on others. This process of sorting out tribal members, determining which were capable of immersion into American life and which needed greater nurturing, led to the establishment of formal guidelines. These guidelines, at least on the surface, appeared designed to elicit information about an individual's background and abilities.[4] Once documented and detailed, this information was to help federal officials categorize tribal members as either competent or incompetent.

The perfection of the allotment process was achieved piecemeal, through a number of laws and amendments. In 1904, the Department of the Interior was granted powers to liquidate Indian land and supervise land sales by issuing fee patents to certain Indians who applied to gain control over their own affairs. For those who applied and received permission from the government, the trust status on an allotment could be curtailed. Tribal members were required on applications to include their age, sex, blood quantum, and business experience, thereby reducing the chance that they might be given too much freedom too quickly.[5] This 1904 legislation was followed in 1906 by the passage of the Burke Act and McCumber Amendment, laws that drew firmer distinctions between those Indians who were deemed capable of mastering their destinies and those who were not.

Under the Burke Act, citizenship for Indians was postponed until the end of the twenty-five year trust period. But the government was allowed to grant citizenship to those Indians who applied for, and were deemed worthy of, a "competency certificate" in advance of the expiration of the trust period. Determining competency could be delicate work. In subsequent modifications to the Burke Act, fee patents were, at first, granted only to tribal members who applied for them, a somewhat self-selecting system. In addition, individual interviews were conducted with prospective allottees. During a competency interview, level of education, fluency in written and spoken English, experience in any form of business, and other relevant circumstances of a tribal member's life were taken into account. Under the McCumber Amendment, which dealt with land sales, restrictions were reinforced upon those tribal members who were deemed incompetent. Incompetents were not allowed to sell their land for the duration of the trust period, and even at the expiration of that time any sale or lease of land was to be supervised by the government.[6]

The application of restrictions on land was hardly a welcome development for all in the Five Tribes. For those wishing to sell or lease their land, to have flexibility in their business dealings, the proposed limits appeared burdensome, even insultingly discriminatory. Tribal members were among the first to point to a double standard in that they were granted citizenship but not the rights of citizens. While conservatives among the tribes favored restrictions, viewing them as the only way to preserve stability in the midst of the tumult caused by allotment, progressives asked that they not be subjected to such protections. Robert Owen, an affluent Cherokee and a member of Oklahoma's first congressional delegation, was among those who maintained that federal restrictions levied an unwanted and unfair disadvantage on tribal members.[7]

Placing land under trust status, while designed to prevent its alienation, effectively deprived tribal members of the ability to enter the mainstream. While it could not be taxed, restricted land could also not be used as collateral to secure loans in order to make necessary improvements, and it could not be sold. Restrictions, it appeared, confined tribal members to subsistence farming or minimal leasing fees and thus limited opportunity. According to one historian, Owen, in a series of speeches before federal officials and various Oklahoma organizations, "pointed to the absurdity of a policy that would protect *him* against exploitation. He supposed he could secure the removal of his own restrictions if he would consent to humiliate himself and prove his competence to some underpaid clerk in the Interior Department" [emphasis in original].[8]

Yet Owen and other progressive-minded tribal members did agree with the federal government's assessment that there existed varying levels of ability among tribal members. In Owen's words, there were many tribal members who were "helpless." These "defectives," as he called them, were akin to "children." Accordingly, it was proper to offer protection for these unfortunate tribal members, to acclimate them slowly to the ways of life in American society. Clearly, these were not the same type of tribal members as those who were prepared to enter the mainstream and become productive citizens.[9]

Running through the government's formulations for extending or lifting protections was the recognition of inerasable, fundamental differences among tribal members. As part of its duties, the Dawes Commission compiled official census lists of the tribal populations in order that allotments might be distributed fairly. These rolls were divided into three large groupings: members by blood, freedmen, and intermarried whites, with each segment sharing equally in the disbursement of resources. The freedmen had become citizens as a result of treaties dating from 1866, while intermarried whites had long made themselves at home among the tribes. In federal legislation affecting tribal resources, these census lists provided a classification system. Often, freedmen and intermarried whites were granted wholesale exclusion from certain laws and regulations. Those included on the rolls as members by blood, however, were effected by a further categorization: blood quantum. The Dawes Rolls denoted each member's quantum, ranging from "full blood," a term indicating that a person possessed at least three-quarters or more Indian blood, to varying fractional amounts.[10]

Blood quantum subsequently became a crucial factor in legislation governing allotment. While "competency" under the Burke Act was to be calculated through various means, the primary tool used by the government to screen candidates was ancestry. A sliding scale of merit was employed which attributed an Indian's official status to his or her blood quantum. Mixed bloods, those with less than three-quarters Indian blood, and especially those with one-half or less, generally qualified for competency largely on the basis of their ancestry if not their actual ability to cope with the pressures of individual landowning.[11] Ancestry loomed large in the stipulations of the McCumber Amendment as well. This act called for closer supervision of land sales involving full bloods and it instituted tighter restrictions on those same tribal members.

In the world of Indian affairs, blood became an indicator by which officials inferred individual traits; blood was a guide to behavior.[12] Erected upon an older scaffolding of prejudice toward Indians, federal policy echoed popular views. "Many white settlers and officials," one writer notes, "maintained a racial stereotype of Native Americans as wild savages and moral degenerates." "White settlers saw Indians as something less than human."[13] Yet these whites were also attuned to the oftentimes subtle differences that existed between tribal members, and they acknowledged that blood was decisive in creating those differences. So, too, did the federal government. Identifying an Indian as a full blood connoted certain attributes, or the lack thereof. For example, it was commonly understood that full bloods were predisposed to traditional religious ritual, cooperative living in all-Indian communities, and keeping their children close to home as opposed to sending them off to school. Full bloods were usually denied competency on the basis of their ancestry and its supposed influence over their behavior. In later years, guidelines handed down by the Indian Bureau outlined that entire classes of Indians could receive their competency without voluntarily applying for a certificate and in the absence of an individual hearing. The chief criteria employed when granting such wholesale competency was blood quantum.[14]

Ironically, while blood was considered by federal officials to have determinative powers over Indian behavior, those same officials dictated Indian behavior on the basis of their blood. In no small measure, an individual's blood quantum affected his future. Members of the Five Tribes might or might not be entitled to federal protection against grafters depending on their trust status, a status conveyed almost entirely by their ancestry. Similarly, tribal members might be allowed to sell or lease their land and enter into the market economy depending, to a significant extent, upon the proportion of white blood running through their veins regardless of their actual ability to deal with white society. In this way, then, blood categorized tribal members and set forth social boundaries within the Five Tribes. According to one historian, "agreements, regulations, and laws regarding allotment had helped create social classes based on racial definition."[15] The government provided the means to keep full bloods Indians and make mixed bloods whites. In turn, many mixed bloods stressed their whiteness so that they might they be released from the subjugated status of Indian wardship and be granted the chance to pursue their ambitions.

Mixed bloods were already well versed in the language of blood and ancestry. George Washington Grayson, a Creek of mixed parentage who served in tribal leadership throughout the early twentieth century, wrote in his autobiography that as a child he was acutely aware of the emphasis placed upon the differences found in mixed and full bloods. He recounted that during his elementary school years his teachers, both white and Indian, were baffled by his inability to comprehend the lessons set before him. Given his mixed parentage and a physical appearance that lacked Indian traits, they considered that he should be able to perform well. The teachers were less surprised, and less concerned, with the poor abilities of a boy who sat next to Grayson, a full blood who was expected to be a marginal student. As an adult, Grayson was a sympathetic advocate for full blood Creeks, but he nevertheless was keenly sensitive to the differences that existed within the tribe and the attribution of those differences to blood.[16]

Mixed bloods in eastern Oklahoma wanted to ensure that their opportunities for upward mobility would not be hindered on account of their ancestry. Just as conservatives proudly pointed to their lineage as something that an infusion of white blood would degrade, mixed bloods held up their white ancestry as a positive attribute. This was evident in marriage trends and views of marriage. In the late nineteenth century, tribal regulations concerning intermarriage contained class as well as racial dimensions. While conservative tribal members desired generally to stop white men from marrying Indian women, progressives took into consideration the type of whites pursuing intermarriage. In the 1880s, the Cherokees required proof of "good moral character" and the endorsement of a number of Cherokees for those whites desiring to marry into the tribe. Progressives wished to allow only those whites of means, and the good character that presumably accompanied those means, to enter the tribe.[17]

The desire to identify with well-heeled whites prompted progressives to look down upon other whites in the territory. Indeed, they were often scathing in their criticism of whites who did not measure up. Having established governments and schools and other institutions of high quality, tribal members saw many white settlers and squatters as crude folk, as poor material upon which to base future generations. These immigrants might be good enough to labor in an Indian landlord's fields, but they were not suitable for marrying tribal women or exercising property and political rights.[18] Tribal members, writes one scholar, "began to take steps to discourage intermarriage or at least to make sure that those whites who married Indians came from the privileged class."[19] In return, white settlers accorded tribal members important distinctions and grew accustomed to classifying Indians by their racial traits, preferring to marry mixed bloods. On the whole "[m]ixed bloods were always characterized as morally and intellectually superior to full bloods."[20] Whites, thus, approved of and reinforced the often subtle differentiations that existed within the tribes; they granted the idea merit and saw its practical applications.

Progressive-minded tribal members understood and were influenced by racial prejudice. They realized that American society's preoccupation with skin color and the power accorded biology could affect them; although their mixed ancestry was a boon under federal policies, it might be considered a taint or deficiency by the ever larger numbers of whites sweeping westward, men and women less familiar with the customs of Indian Territory. Because of this, they emphasized their white blood and disassociated themselves from Indians who lacked a leavening whiteness and who stirred up troubles that invited persecution. Progressives saw their future as interwoven with white society and they did not want to be associated with elements of the tribal populations that shunned whites.

But the task of promoting their abilities to the exclusion of their less sophisticated tribesmen was a difficult one. Resentment mounted among many whites who considered that Indians received preferential treatment from the federal government in the form of tax-free trust status. White settlers and politicians scorned the fact that "Indians were seemingly rewarded for their laziness with gifts of government food rations, clothing, agricultural implements and seed."[21] Such treatment smacked of Reconstruction Era policies in the South that aided African Americans, a development that was particularly disturbing among the many southern whites arriving in the

West. In Indian Territory, a place dubbed "Little Dixie" because of its high concentration of ex-southerners, even the faintest hint of racial favoritism was bound to stir up negative fanfare. And federal programs to protect Indian lands were hardly a minor feature of life in the eastern portions of Oklahoma. Moreover, eliminating whites' traditional views of Indians as ignorant savages proved impossible, especially when those views were coupled with the often high-profile activities of conservative organizations such as the Snakes.

These were not idle worries. During instances of conservative resistance, progressives feared they would be associated with the troublesome lawbreakers. Such was the case during the Creek Snake troubles of 1900 and 1901, when disgruntled separatists erected a shadow government and took up arms against allotment. In reaction, many mixed bloods grew defensive. During the Snake crises of the early twentieth century, "[t]he majority of Creeks were embarrassed by [Chitto Harjo's] uprising. They had resigned themselves to the imminent changes [associated with allotment] and condemned Chitto Harjo for undertaking a 'hopeless' cause. They wanted United States citizens to understand they were cooperative in spirit."[22]

Progressive embarrassment was grounded in a heartfelt concern for their futures and those of their children. There was no mistaking the anti-Indian sentiment that the Snakes and other groups aroused. Conservative resistance provoked both physical retaliation by whites and damning attitudes toward all Indians. Following separatist troubles in the tribes, neighboring whites and law enforcement personnel routinely swept through tribal communities to restore order. Indians were arrested regardless of their association with the specific groups blamed for the unrest, and homes and crops were indiscriminately destroyed. Progressives fretted that resistance only invited oppression. Although they understood that large numbers of tribal members were not adequately prepared for allotment and assimilation into the mainstream, they argued that defiance was an improper course that had repercussions for all Indian residents.

Progressives had spent decades establishing and burnishing distinctions between themselves and their tribal brethren, contrasting their achievements and plans for the future with those of conservatives. The hysteria produced by the type of "Indians-on-the-warpath" incidents created by the Snakes and other groups threatened the delicate order that existed among the tribes. During times of crisis, when whites called for military aid or, worse, took matters into their own hands, anti-Indian attitudes threatened to tar all tribal members with the same brush of dangerous inferiority. In response, progressives attempted to perpetuate the distinctions among tribal members and to divorce themselves from the stubborn vestiges of "old time" Indian life as personified by men such Chitto Harjo. An important basis for that divorce was blood. Progressives repeatedly pointed out that separatists were full bloods and that their acts were the products of a full-blood mentality. They adopted a patronizing tone when discussing their conservative tribesmen, much as Robert Owen did when he labeled them "children" and "defectives."

Their shortsighted resistance was thus attributed to their deficient lineage and their lack of innate civilization.

These attitudes were further complicated by the uneasy relations both whites and tribal members maintained with African Americans. An incident in 1909, again involving the Snakes in Creek country, revealed the extent to which progressive-minded tribal members were concerned not only about conservative resistance but about the close relationship, real and perceived, between blacks and Indians. The racial dimensions of this event suggest the extent to which race had come to play a vital role in tribal affairs, in the divisions among tribal members and relations between Indians and others outside the tribes.

Beginning in 1907, a small encampment of blacks adjacent to Hickory Ground grew to several hundred people. With few prospects, these black migrants and former freedmen had requested permission to stay temporarily. At the same time, the camp heightened racial tensions in the vicinity. Whites and many Creeks were unhappy with the black presence. Matters worsened when several burglaries, most involving food, were blamed on the blacks. In the atmosphere of resentment and suspicion that enveloped the town, the theft of meat from a white farmer's smokehouse in March 1909 triggered the simmering anti-black sentiment and led to bloodshed. White residents assembled a posse under the supervision of the sheriff, and armed men began interrogating Creeks and blacks at Hickory Ground. Run-ins with apprehensive Snakes touched off gunfire and, later, a siege of a cabin where Chitto Harjo was living. After two days fighting, several people were dead and others wounded. Chitto Harjo fled under cover of night, but the white vigilantes, inflamed by pent-up hatred of blacks and their Creek friends, attacked the town. They razed homes and destroyed crops. Numerous Creeks were jailed, some of whom had little or nothing to do with the Snakes, including an 88-year old man and 13-year old boy.[23]

White Oklahomans did not quibble in respect to their anti-black sentiments. As in the South and other regions of the United States, blacks received ill-treatment in Oklahoma. Indeed, prior to statehood, during their initial forays westward into Kansas and Oklahoma Territory, blacks had found it prudent and necessary to establish towns separate from white communities.[24] These outcasts did not reap the bounty of freedom and opportunity associated with westering; instead they endured the powerful reach of racial discrimination. In time, anti-black attitudes only worsened. Particularly in the southeastern portions of Oklahoma, the home both of the Five Tribes and many ex-southerners, blacks endured monumental prejudice. They lived in a society as intolerant as the one they had left behind in the post-Civil War South, and they endured widespread and debilitating discrimination. Although blacks were granted citizenship under Oklahoma's constitution, something achieved despite the vehement protests of the constitutional convention and only after threats by President Theodore Roosevelt to veto the state's enabling act, they faced crippling restrictions soonafter. In the first legislative session following statehood, blacks were banned from public schools and a host of public places and spaces.[25]

The curtailment of the rights of blacks was further evident in the repeated violent attacks administered by white Oklahomans and law enforcement organizations. There were frequent lynchings and other racial clashes. Throughout the late nineteenth and early twentieth centuries, it appeared that whenever a town reached a certain point of development it went through a phase of racial cleansing, with blacks fleeing into the night while their homes were set ablaze.[26] Indeed, the troubles at Hickory Ground grew out of the racial violence that was endemic to Oklahoma. Just two years prior to the Smoked Meat Rebellion, a vicious riot in Henryetta, located six miles from Hickory Ground, resulted in the expulsion of blacks from the town. Touched off by a shooting incident, the town's residents turned on their black neighbors with a fury. Homes were burned and pitched battles exploded through the streets. So sensitive were racial relations that from 1907 to 1931 blacks were not allowed to stay overnight in Henryetta. Driven out of town, many arrived at Hickory Ground seeking refuge. There they lay the foundation for the encampment that in two years would soon grow to significant size and serve to incite new troubles.[27]

For progressive tribal members, the link between Indians and blacks, even if it was a link that existed primarily as a popular fiction embraced by whites, was a dangerous one. Their experiences showed quite clearly how Americans treated blacks as well as other peoples considered "inferior." They themselves had historically been the subjects of race-based policies, as during their removal from the Southeast, and they exhibited a long-held fear of being thrown together with other degraded peoples. In the early twentieth century, as a rising tide of racial violence was cresting, progressives were eager demonstrate their fitness to assume the opportunities held out to them in Oklahoma. They struggled to downplay the negative stereotypes that haunted Indians, including any connection with blacks.

The mistreatment of Oklahoma's blacks took place in plain sight of the Five Tribes, and tribal members possessed an intimate understanding of racial tensions and its political undertones. In 1910 an article re-printed in the *Tahlequah Arrow*, published in Cherokee country, explained why it was important to separate blacks from Indians. Although the article, entitled "Will Swat the Negro," was partisan in nature and an attempt to discredit Republicans, it nonetheless put forth powerful and widely accepted themes. "The Indians, as a people, are a proud race. They have a right to be. The efforts of the republicans *to class them along with the negroes* now will be of no avail" [emphasis mine]. A grandfather clause excluding blacks from various rights in Oklahoma "will be his [the Indian's] first chance to swat the negro since the time the negroes grabbed the Indians' land, and he is going to swat him good and hard."[28] But such reports did not have to manufacture animosity between Indians and blacks. In fact, the history of Indian-black relations among the tribes was itself fraught with tension and myriad troubles.

In the late eighteenth and early nineteenth centuries, tribal members owned black slaves, many of whom the Indians brought west with them during removal. Slaves toiled in the fields of the tribal elite, serving much as others did on Southern plantations. Other structures of slavery similarly prevailed in the tribal nations. Miscegenation laws had a long history in the tribes; a Cherokee law from 1839 prohibited marriage between Indians and blacks under penalty of physical punishment and monetary fine.[29] Under the Cherokee constitution, citizenship in the tribal nation was granted to the descendants of Cherokee men by all free women except those of the African race.[30] Conditions did not improve following the Civil War. Indeed, Indian Territory witnessed nagging problems involving relations between Indians and the freedmen. Having absorbed many of the attitudes that whites exhibited toward blacks, tribal members were not happy with the federal government's intervention in racial affairs at the close of the war. In treaties drafted in 1866, the government demanded that tribal freedmen be granted citizenship in each of the tribal nations. As citizens, they would possess property and political rights.

But relations between Indians and blacks were so strained that the government was forced to amend its treaty demands. The Choctaws and Chickasaws, the most southerly of the tribes, received special provisions in their treaties. If popular referendums were passed to evict the freedmen from the tribe, the United States offered to pay for the freedmen's removal. While the Choctaws, with great reluctance, relented and extended rights to the freedmen, the Chickasaws stubbornly refused. The tribe passed the referendum and demanded that the federal government make good on its promise, which it never did.[31] In 1876, when the issue of accepting the freedmen into the Chickasaw nation was again raised, under pressure by the federal government, the tribe's governor, Benjamin Overton, raised the alarm. "If you do [admit the freedmen], you sign the death-warrant of your nationality with your own hands; for the negroes will be the wedge with which our country will be rent asunder and opened up to the whites; and then the grand scheme so artfully devised by the treaty of 1866, will have been effected, and the ends of the conspirators attained."[32] The Chickasaw freedmen, subjected to mistreatment throughout the reconstruction period, became, in the words of one writer, truly a "people without a country."[33]

The other tribes proved similarly resistant to black equality. Although they formally extended rights of citizenship to their freedmen, such concessions were proffered reluctantly. The latter half of the nineteenth century was crowded with incidents of racial conflict between Indians and blacks. The Cherokees endeavored to rid their nation of blacks and numerous violent episodes between the races were recorded. According to one scholar, the Cherokees' abuse of their freedmen, as well as black migrants from the South, constituted a "pogrom."[34] Such animosity also spilled over tribal lines. In 1880, groups of Cherokees, complaining that freedmen from the neighboring Creek nation were rustling cattle on Cherokee lands, captured two suspects and lynched them.[35]

The Seminoles, credited by some with more egalitarian treatment of the freedmen, effectively denied many rights to those of African American descent. The anthropologist Rebecca Bateman notes that the level of animosity between Indians and freedmen by the early twentieth century had grown intractable.[36] In a cruel irony, Seminole freedmen, who considered themselves to be Seminole and who regarded other blacks as an entirely different ethnic group, were forced to form an alliance with "state negroes" emigrating westward from the South as a result of the Seminoles' insistent attempts to exclude them from participation in tribal affairs. Bateman credits Oklahoma society's willingness to recognize Indians as "white" with spurring anti-black attitudes among Seminole tribal members. The process of racial segregation erected by Oklahoma lawmakers, she observes, made a large impression upon the Seminoles. They saw that they, as a group, possessed an advantage that their freedmen, who it was understood had been thrust upon them by treaty in 1866, did not.[37] This was particularly the case among mixed blood tribal members who wanted to keep freedmen affairs separate from those of the rest of the tribe.

The separation between blacks and Indians carried over into the private lives of the tribes. Despite the popular perception that blacks and Indians intermarried frequently, the so-called "myth of mixture," the actual figures were quite low.[38] One historian, working on a sample derived from the Creek nation in 1900, found no cases of mixed blood-black intermarriage. In addition, the instances of full blood-black intermarriage made up less than two percent of all marriages. Among the Cherokees, intermarriage was even rarer. There were, in 1900, no cases of mixed blood-black marriage and only 1.1 percent of all marriages in the tribe involved unions of full bloods and blacks.[39] In those cases where intermarriage did occur, it was cause for hostility. The Chickasaws repudiated marriages, common law and otherwise, between blacks and Indians by granting them no official recognition. This was in contrast with the practice of the tribe to recognize marriages between whites and Indians. An 1885 law mandated a punishment of fifty lashes for tribal members who married blacks. As early as the 1850s, in fact, the Chickasaws outlawed mixed marriages and subjected offenders to banishment and heavy fines.[40]

One reason for the cold attitude toward blacks and for black-Indian marriages was a widely-held belief among the tribes that the number of blacks increased far more quickly than that of Indians. While full bloods grew increasingly concerned that their pure lineage was imperiled, mixed bloods were concerned with the potential taint that blacks posed for tribal members. In explaining their refusal to grant their ex-slaves political rights, the Chickasaws had maintained in the late nineteenth century that "the freedmen and colored immigrants constitute so large a part of the Chickasaw nation and increase in number so rapidly that they must soon outnumber the Chickasaws, and, if invested with the elective franchise, will be able to take possession of the government, and ultimately to deprive the Chickasaw people of their government and country."[41] The Chickasaw governor, Douglas Johnston, explained the racial calculus of the tribe. In arguing against inclusion of the freedmen on the tribal rolls, he stated: "it will be but a few generations until the full blood Indian will be no more, but as the Indian citizen vanishes, the 'Negro Chickasaw,' if such he is made by Congress, will multiply, and the time will not be far distant, if this iniquity is visited upon us, when the name of Chickasaw will carry with it [o]pprobrium and reproach instead of honor."[42]

The general resentment against blacks was aggravated as a result of the Dawes Commission's work in the region. The allotment of tribal lands had necessarily dealt with the issue of freedmen rights and freedmen enrollment. The tribes, as they had done in the nineteenth century in limiting the claims of freedmen for full political and property rights, attempted to bar the freedmen from an equal share of land under allotment. The Chickasaws, for example, after refusing to admit the freedmen into the tribes in the wake of the 1866 treaty, acted to ensure that blacks be prevented from being enrolled on the "blood" rolls of the Dawes Commission's census. They also hoped to keep the children of black-Indian unions from the rolls, thereby lessening the chances that membership might include blacks. The "Chickasaw mixed-blood elite," according to one scholar, "insisted that the Chickasaws were for the most part free of African blood" and thus succeeded in keeping those with black ancestry off the rolls.[43]

Many tribal members expressed their dissatisfaction with the federal government's intention to distribute land and mineral rights to freedmen. Chitto Harjo himself had questioned the right of the freedmen to stake claim against Creek lands. Three years prior to the Smoked Meat Rebellion, Chitto Harjo stated:

> I hear the Government is cutting up my land and giving it away to black people. I want to know if this is so. . . . These black people, who are they? They are negroes that came in here as slaves. They have no right to this land. It was never given to them. . . . Then can it be that the Government is giving it—my land—to the negro?[44]

A fellow Creek Snake, Eufaula Harjo, expressed a similar bitterness toward the "negroes . . . that have taken all the land, and there is nothing left for the full-blood Indian at all."[45]

While conservative sentiment toward blacks was hardly as accepting as whites alleged, or as some scholars have portrayed, progressives were more forceful in their statements regarding blacks.[46] They were also more likely to spot the political advantage in the prevailing anti-black attitudes. In one newspaper editorial, the writer noted that influential tribal members weighed their political support upon the candidates' views toward blacks. "The average Indian," the editorial maintained, "especially of that class which control political matters of his nation, considers himself as far above the negro socially as does the white man."[47] The noted Creek writer Alex Posey similarly advanced these ideas, and, according to his biographer, "expressed a virulent racism aimed at people of African descent." "To [Posey], progress did not include either political power or social equality for blacks."[48]

As mixed bloods sought to enter white society, their anti-black racism became more pronounced. Progressives saw the usefulness and importance of convincing whites that they were not akin to blacks. They supported efforts at the constitutional convention to grant Indians legal equality with whites. In a provision submitted by Gov. Charles Haskell, all persons who were not identified as being of African-American descent were regarded as being white. According to the *Muskogee Times-Democrat*, Haskell's measure "provide[d] that there shall never be any distinction between whites and Indians."[49]

But tribal members were not satisfied with keeping the Indian and black races apart in actuality. They also recognized the need to combat the *perception* that the two groups were joined together. It was commonly accepted among many whites that the tribes, through close association with blacks over a period of decades, had become "polluted." Although intermarriage rates were in fact low, whites considered that extensive Indian and black intermingling had taken place and had produced a hybrid race. A magazine article in 1907 argued that among the Seminoles and Creeks, intermarriage with blacks was rampant.[50] Elsewhere, another writer maintained that "intermarriage became so common, so that now (I have it on the best authority) there is not a Seminole family that is entirely free from negro blood; and there are but three Creek families, some make it two, that are of pure blood."[51] It was this perception that so troubled many tribal members and prompted Chickasaw Governor Douglas Johnston to emphasize that Indians viewed any possible classification of Indians alongside blacks as intolerable, "and the white race under similar condition would have the same feeling."[52]

The anti-black attitudes of progressive tribal members sprang from the same source as their exertions to divide themselves from conservative tribesmen. They were both part of a shift toward racial identification. Previously, divisions within the tribes were predicated upon class standards. But by the early twentieth century, race-based standards overlay class distinctions and widened the gulf among tribal members. For full bloods, their status as restricted Indians was an extension of their ancestry. For mixed bloods, their status as competent Indians was similarly the result of their parentage. Both groups, albeit in different ways and for different purposes, doted on biology; they considered a person's blood a telling fact.

* * *

Ironically, while the whiteness of mixed bloods promised to usher them into the mainstream, the destruction of tribal institutions as mandated by the Dawes Agreements eliminated an asset that had served them well. Tribal institutions allowed progressives to succeed, to build up impressive school systems, participate in constitutional governments, and embark on careers in business and the professions. Membership in the sovereign Indian nations provided the basis for much accomplishment. A person belonging to a tribe was granted political and property rights, and hence significant material advantages, that non-members did not possess. More importantly, progressive members had been quite successful in gaining power within each of the Five Tribes and gearing tribal policies toward their own success. With statehood, however, they stressed their white ancestry at the expense of their Indianness. In this, they lost a valuable asset; being Indian had once served them well. To be Indian and vested in a sovereign nation was to occupy a privileged position. To be Indian, no matter how white, in white society was, progressives would discover, something else entirely.

FIVE
Hardship and Decline

Assigned as a field matron with the Indian Bureau, Susie Ellis was touring a number of remote Five Tribes communities in 1917 when she reported the distressing realities of Indian life in rural Oklahoma.

> My work is hard, as the Country is very rough but I fear I am unable to impress upon you the pitiful conditions I find in some of the Full Blood huts and homes in the isolated parts of the mountianous Country. . . . Of course, it is caused by the peculiarity and ignorance of the Indian, also to poverty and [unsanitary] conditions. I have gone into homes where as many as five and sometimes eight or ten lived in a one room log hut, that leaks. No ventilation and as many dogs as there are people. And dogs with mange.[1]

Of particular importance to Ellis, the "greatest of all needs," was improved care for women and children. To illustrate her point, she recounted a grisly tale. "Last week, I walked into a home, the home of Leslie Beaver, whose child three weeks old was dead, died of a fever. The pitiful sight of this mother over the crude coffin with rags placed around the tiny wasted form to hold it in shape, to me was heart rending."[2]

In the years following Oklahoma statehood, the overall conditions of tribal members, especially conservatives, deteriorated rapidly. In the wake of allotment, these rural poor lost land at an astounding rate. Many sold off their parcels and netted scant reward. Others were preyed upon by grafters who descended upon the uneducated and the defenseless like locusts, stripping them of their resources, however meager. Still other tribal members simply could not contend with the changing nature of Oklahoma's economy. Small and subsistence farmers increasingly occupied a marginal position in a western economy where cash crops and mechanization increasingly occupied an important and profitable position. Mired in poverty and ignorance, conservative tribal members often lived in unhealthful conditions, and disease was a chronic visitor among their settlements. Food was often in short supply and there was little money to supplement subsistence farming.

These misfortunes painfully exposed the shortcomings of the conservatives' previous strategies of separatism and resistance. Attempts to live apart from white society and preserve older customs and ceremonies were failing, often spectacularly. Resistance did not bring about the withdrawal of white settlers from Oklahoma or the persistence of tribal institutions. To the contrary, defiance provoked oppression and violence. Consequently, tribal members relied increasingly upon the federal government for direct aid as well as protection against the demands of whites and increasingly confident state and local authorities.

The effect of these developments had a curious impact upon tribal relations. Indian victimization at the hands of whites served to revive links between tribal members of varied background and political leanings. The plight of conservatives, so awful in many circumstances, aroused the sympathy of their progressive tribesmen. Tragedy lessened older factional animosities. While progressives had previously attempted to remain apart from their tribal opponents and critics, they now intervened on their behalf. Using the Indian bureau, they tried to relieve suffering. This did not mean that they abandoned their own assimilationist course and their own distinct ambitions, but it hinted at the bonds that still existed between tribal members, bonds not eradicated by the Dawes reforms and citizenship in Oklahoma.

* * *

Throughout the first decade of statehood, tribal members faced harsh poverty and an accompanying litany of problems. Many tribal settlements, especially those inhabited by conservatives and supporters of separatist movements, were isolated. They were places where families worked small patches of ground, lived on what they grew. These tribal members generally possessed few tools in good repair and lacked adequate seed to carry over from harvest to planting. Their clothes were threadbare, and their homes were drafty in winter and ill-ventilated in summer. For assistance, conservative tribal members most often depended on churches, neighbors, or the government in times of drought or other difficulty.[3]

At various times following Oklahoma statehood, area residents or observers passing through the region alerted the Indian Bureau or the Department of the Interior to the dire conditions in which conservatives lived. Writing from Eufaula, OK., in November 1907, E. W. Loomis, who resided in Louisville, Kentucky, wrote President Theodore Roosevelt of hardships endured at harvest time. "I write to you in behalf of the Indians of the

Hardship and Decline 53

Cherokee and Choctaw Nation. . . . They have had short crops and the full bloods are to[o] proud to beg and assistance will have to reach them at once or there will be terrible suffering."[4]

In 1913, the Commissioner of Indian Affairs received a letter informing him that numerous Cherokees were on the verge of starvation. A punishing drought combined with crushing poverty imperiled subsistence farmers, many of whom were members of the Keetoowah Society. Confirmation of the distressing specter hanging over the Cherokees was soon forwarded by field agents who were dispatched to learn the extent of the crisis. One field agent reported that while the region's cotton crop was thriving, hoped-for food crops were "almost universally a failure throughout the district." These circumstances, it was noted, not only jeopardized the immediate survival of the tribal members, but threatened their hold on what land they still retained. "There is always a great demand on the part of the retrograde Indians of the district for cash payments, whether land sale, equalization, or other moneys, and I rather expect this demand to be more acute this winter than it has heretofore been," a field clerk wrote.[5] "In some cases (no doubt in a good many cases) it will be necessary to grant these appeals" allowing tribal members to sell their allotments.[6] In the event the government allowed tribal members to sell their lands in order to survive, it was understood that they would likely never be able to re-purchase it. What appeared a seasonal crisis, one caused by the vagaries of the weather, thus contained the seeds of future disaster.

Again, in early 1917, letters reached bureau officials describing crop failures and attendant miseries. G. W. Clark, a self-described "3 quarters Cherokee woman" of 71 years, wrote that the corn crop amounted to a total loss producing "this eleventh hour of suffering."[7] For the full blood Cherokees living in the eastern hills, tough times only promised to get worse. "They have no corn to live on . . . and they have no way of making money to buy even meager livelihood." "I have lived among them for all these years," Clark added, and "I never saw them in such destitute for food & clothing. Even during the Civil War times was not so hard."[8] Another writer, Ben F. Smith, posted a letter to O. K. Chandler, a bureau field clerk, in January 1917 and related to him the doings of "about a half dozen or more [Chickasaw] families living in this community (Steeley, Oklahoma) that ought to be done something for them. . . . [T]hey are needing cloths and shoes worse than anything else, most of these folks are making cross ties and hauling them to Strang[, OK.] and making enough money for 'grub.'"[9] Chandler forwarded Smith's letter along with his own request that some old police uniforms be distributed to the destitute Indians.

Similar cases abounded among each of the Five Tribes. As historians have documented, white settlement, landlessness, and severe social dislocation pushed many tribal members into a downward spiral of unrelenting hardship.[10] For those who vainly tried to retain local autonomy and ward off allotment, the changes sweeping the region posed grave challenges. In the face of these difficulties, tribal members fared badly in their attempt to remain outside of the market-driven economy. Although conservative leaders uttered defiant rhetoric and urged their followers to refuse allotment certificates and otherwise avoid contact with white society, it was growing impossible to ignore the pressures deriving from large scale economic upheaval in the region.[11]

In eastern Oklahoma, white settlers had gradually overwhelmed the Indian population in the late nineteenth century. As they did so, they established new systems of land use and ownership. Upon invading Little Dixie, ex-southerners settled on a patchwork of smaller farms. There they grew a variety of crops including wheat, corn, and cotton, and raised livestock. They were subsistence farmers who also produced a small surplus for the market.[12] During the late nineteenth century, these settlers demanded access to Indian land in order to fulfill their desire for places of their own, and some worked as tenants for Indian landlords.

By the early twentieth century, however, the population of the southern and eastern portions of Oklahoma soared. From 1900 to 1910, the number of residents in the former Indian Territory doubled. With increased demand, land prices also rose. As a result, remaining Indian-owned lands were even more greatly coveted. At the same time, a new and unhappy realization began to dawn on those arriving in the region to pursue their dreams: the price of independence was moving ever higher and out of their reach. New economic trends began to exert themselves in agriculture. A small plot of land was priced beyond the means of the family farmer. What was more, even if land was acquired, the investment required to make it profitable required a more intense style of farming. To make the land yield greater returns, farmers invested in heavy machinery. Machinery was most efficiently utilized on larger tracts and some farmers began to acquire sizable holdings by stitching together smaller parcels. In the event, bigger operations with substantial investments in land and technology and the accompanying imperative to turn a profit sprang up in eastern Oklahoma.[13]

As the region became enmeshed within domestic and international markets, Oklahoma farmers turned to crops that offered the best returns. An emphasis on specialization crowded out diversified farming and subsistence.[14] A cash crop economy based upon cotton emerged. "Cotton prices began to rise during the first decade of the twentieth century," one writer observes, "and Oklahoma acreage in this crop almost tripled. Cotton soon became the staple crop in the southern half of the state."[15] One of the negative effects of this switch to cash crops soon became obvious. Whereas tribal members previously planted various crops on their lands, the rising proportion of acreage devoted to cotton left them vulnerable not only to market forces, but also to deprivation. Indeed, as with the Keetoowah farmers in Cherokee country in 1913 referred to above, devastation to their corn crop left them starving while their cotton crop was growing nicely but not ready to be harvested and converted into cash.

Although the rural economy expanded in the state's first years, part of what one writer calls Oklahoma's "Age of Expansion," it was decidedly not a time of "universal prosperity."[16] Small farmers, both Indian and non-Indian, were unable to preserve their station and many fell into tenancy. The rise of tenancy revealed a darker side of the state's development. "Ideally, tenancy was supposed to serve as a rung on the 'agricultural ladder,' whereby a poor man could gradually rise to full ownership of a farm. But in Oklahoma a permanent tenant class had developed."[17] In the seven eastern counties that formerly comprised Indian Territory, roughly sixty percent of the population by World War I remained tenant farmers. Disparities in social and economic position prompted political unrest, and the Socialist Party attracted an especially vocal and popular following by the 1910s. Oklahoma also displayed high rates of farm union and organized tenant activity.[18] For some writers, these developments offer a commentary on the state's colonialist status within the larger nation. As a producer of agricultural goods, and lacking a mature manufacturing base, Oklahoma had become "tributary and subordinate to older, richer regions."[19]

But while many whites endured hardship, Indians suffered a dramatic and tragic decline. Federal policies often restricted Indians from using their lands as collateral to obtain loans and whites were loathe to assist them. Indians were deemed bad risks and they were resented for the "coddling" they supposedly received from the government. When capital was available, it came with an extraordinarily high price tag and numerous conditions; the usurious lending practices of Oklahoma banks in the early twentieth century were legendary. On the whole, despite gaining control over their allotments, Indians were badly positioned to reap the rewards of the modern agricultural economy.[20] Individual land ownership, which non-Indians held out as the greatest asset in modern society, was not, as it turned out, sufficient to ensure safety and comfort. The assurances of the Dawes Commission, that the Indian's claim to Americanness flowed from private ownership of a plot of land, rang hollow. Subsistence was no longer an option with the advent of a competitive market economy. What good was the land if a person could not pay for improvements? And what of the constant threats of taxation? Oklahoma authorities were relentless in their bid to place Indian lands on the tax rolls. Indians needed credit and cash, as did other farmers. Large farmers with access to capital were the ones who took advantage of cotton growing, not the subsistence farmers or tenants who labored on others' lands.

Even during robust economic times, as during World War I, Indians rarely benefited. The rising demand for agricultural products that accompanied America's mobilization for war in the late 1910s ostensibly placed Indians in an advantageous position during the national emergency. When prices rose, tribal members eyed the chance to reap a windfall. They even received significant assistance from the federal government, which initiated an agricultural campaign to convert Indians into independent farmers and to put idle Indian fields under the plow.[21] But despite boasts of the program's success, the rise in wartime production and prices primarily benefited white farmers who were able to buy machinery and lease or purchase more land. The large scale economic changes already re-making Oklahoma accelerated during the war, profiting whites.

Doubtless, Indian industriousness was put to better advantage during the war, but members of the Five Tribes nonetheless lagged behind other Oklahomans. William Neighbors, a tribal member living in Sapulpa, OK., wrote to the Secretary of the Interior during the war to complain that Indians tended to be penalized, not rewarded, for their wardship status.

> [N]ow we the Indians I speak of, I dont mean all but the majority as a rule[,] have the land and no money[.] [O]ur land is restricted[,] cant borrow or sell[.] [T]hose that are not restricted have sold or borrowed to the full limit. . . . If I had the money I could buy different kinds of seed to plant beans . . . sweet potatoes and I have to buy feed and it all takes money.[22]

Caught between a federal government offering assistance and a marketplace that discriminated against their restricted status, tribal members could only watch as the wartime boom largely passed them by.

Writing from the town of Ada, OK., B. V. Hampton was more critical. Addressing Congressman C. D. Carter in April of 1917, Hampton wrote "in regard to the Indians who have to go farming to raise food stuff for warring nation. Understand I am not [opposed] to such plan. But if we are going to farm our own lands we ought to have a chance as well as all other farmers. A great many of our Indians haven[']t the means to make a crop for this purpose not even for them selves." To enable Indians to take their place alongside Oklahoma's other farmers, Hampton maintained, "[t]he government will have to go to great expense to supplies. A great many [Indians] have no teams, no farming impliment." He offered himself as an example of how Indians were expected to make their way in the growing commercial society with no leverage. "Take my case for instance. I have 140 acres. But I have rented it all to tenant farmers for this year. Also I have neither teams nor farming impliments." Even endowed tribal members, Hampton made clear, faced grave difficulties.[23]

While wartime policies proved inadequate, the war's end spelled greater trouble. When peace arrived, farmers across the state coped with a brief but acute depression. Shrinking prices prompted many to increase the amount of land under cultivation to counter the setback. These fluctuations, in turn, furthered the trends already transforming the rural economy. Whereas increased mechanization and large agricultural operations existed prior to World War I, economic pressures heightened the reliance upon cash crops and heavy machinery, spawning a vicious cycle of overproduction and dropping prices. Yet in the face of these large scale structural changes, the Indian bureau continued to foist an outmoded response upon tribal members. The thrust of policy, even as large landholding and mechanization pointed to

Hardship and Decline

future competitive practices, was to encourage tribal members to be satisfied with independent toil on small allotments. Yeomanry, a concept of the eighteenth century, was the prescribed solution for a rapidly changing twentieth century economy.[24]

Indians were not only subjected to the vagaries of economic change, but were the targets of a cruel campaign of corruption and theft by white Oklahomans. Many tribal members were little prepared for the orgy of acquisition, and the crime it spawned, that preceded formal statehood and which proceeded, without interruption, through the early twentieth century. As Angie Debo has eloquently and exhaustively documented, chaos and rampant white greed descended upon the Five Tribes in the wake of Oklahoma statehood.[25]

Grafters were relentlessly creative in their schemes to defraud tribal members. They pursued their prey despite federal protections and the efforts of government to discourage them. One historian has written that

> the entire Five Tribes area was dominated by a vast criminal conspiracy to wrest a great and rich domain from its owners. In the process of leasing restricted land or purchasing the unrestricted land from ignorant adults or ignorant or corrupt guardians the allottee was overreached by every possible sharp practice or criminal action.[26]

Merchants often lent allottees money with the sole purpose of entrapping them in indebtedness. Real estate speculators then drew up sale contracts and initiated contact with the indebted allotee. Attorneys routinely filled out blank applications for lifting restrictions for fees of $50 to $100. Although these sales frequently became hopelessly entangled in the courts, they nonetheless succeeded in mystifying tribal members, many of whom gave up their fight to keep the land. Even as the federal government tried to sort out the convoluted series of cases that arose from bogus land sales, the sheer volume was often too great.[27]

Other grafters employed strongarm tactics to obtain Indian signatures on fraudulent sale contracts. Kidnapping was routine. Tribal members were often spirited across state lines to Missouri or Arkansas where they were held until they agreed to forfeit claim to their rightful allotments. The use of alcohol to "persuade" tribal members to sign away their allotments was also a popular grafting method. On occasion, alcohol and kidnapping were used in conjunction. Indians might find themselves being offered whiskey only to wake up hours later in another state confronted by men with guns demanding allotments.[28]

Murder, too, was employed to obtain land from allottees and their children. Speculators were known to keep "Birthday books," which listed the dates when minors were to reach their majority and thus gain rights over their lands. There were not a few cases of young tribal members on the eve of their birthdays who were served doctored alcohol, forced to sign away their allotments, and then left dead on a roadside. Autopsies were almost never performed in such cases, since drunken Indians did not merit careful investigation to determine the causes of their deaths and the potential motives of those involved.[29]

Tribal members entered, or were forced, into dealings with grafters for various reasons. They were frequently ignorant of the intentions of those who befriended them and offered them loans of money. Or they were in such dire need of cash that an undervalued purchase price could not be refused. If they agreed to lease or sell portions of their land, they frequently did not know how much land they agreed to part with. Nor did they understand the ramifications of their actions. The market in allotments was so active and so profitable that tribal members were lost in the swirl of exchanges. One writer has noted that 72,000 members of the Five Tribes were landless by the 1930s, and that total tribal holdings had declined from 15 million acres at the turn of the century to 1.5 million.[30]

Corruption in Oklahoma assumed its most refined, systematic, and ghastly heights when the targets were tribal minors. By the 1910s, forty county judges had jurisdiction over sixty thousand minors whose land was worth an estimated $130 million, as well as an "oil valuation of $25 million."[31] Courthouse "rings" preyed upon tribal minors and legal incompetents. Judges frequently appointed cronies as guardians, and it was not unusual for a single guardian to be placed in charge of the affairs of a number of Indian minors. Guardianship abuses were such that it was rare for these actions ever to be questioned and corrected. Attorneys for the Choctaws, in two rare victories for tribal members, prevented individuals from excessive abuse of the guardianship laws. In one case, a man narrowly lost his bid to "secure the guardianship of 350 children from one county court and the approval of the sale of timber from the whole 350 allotments."[32] In another instance, a man nearly obtained power over "140 children in order to sell inherited land scattered throughout the Choctaw and Chickasaw nations," before attorneys employed by the tribe prevented such abuses.[33]

But many more cases of fraud and mismanagement eluded the authorities. Heirships were hotly contested by attorneys, guardians, and speculators, but generally not to the benefit of the heirs themselves. Much Indian land was diverted through the probate courts and was never handed down to family members; it was instead snapped up by land traders who obtained illegal access to it. Land, while the principal objective of various guardian schemes, was not the only cause for extortion. Given their virtually complete control over the assets of their charges, guardians were able to charge exorbitant fees for their expertise and advice on behalf of their wards. There were numerous examples of guardians paying themselves handsomely from the Indian money accounts of tribal members while leaving the tribal members themselves half-starved, ill-clothed, and dependent upon relatives for sustenance and shelter. One report, furnished by the Creek national attorney, demonstrated that guardians charged an average of twenty percent of an Indian's assets to administer the estate. This compared to just over two percent to handle matters related to white wards.[34]

While corruption raged on, Oklahoma officialdom pursued Indian assets in more conventional, that is to say legal and political, ways. Irritated by the continuing influence of the federal government in Indian affairs, state officials sought opportunities to limit federal control while enhancing their own. If tribal members fared poorly under the Dawes Agreements in the decade preceding statehood in 1907, they endured greater distress under the increasingly confident and powerful control of their bridegroom, the state of Oklahoma.

In 1908, Oklahoma's congressional delegation won an important victory in the battle between state and federal authorities for control over Indians and Indian lands. Federal legislation was passed that sped the removal of restrictions protecting allotments and opened vast new tracts to both taxation and sale.[35] According to the law, which remained in force for the next two decades, tribal members were divided into three categories for purposes of determining the applicability of continued federal oversight of their affairs. The unrestricted population consisted of tribal members who were white, freedmen, or mixed bloods of one-half or less Indian blood. These individuals were free to sell their lands outright. They were also subject to taxation, a measure greeted gleefully by white Oklahomans and officials. Mixed bloods of one-half to three-quarters Indian blood were allowed to sell their surplus acreage but not their homesteads; their homestead lands could not be taxed as long as they remained in federal trust. And those of three-quarters or more Indian blood retained restrictions on all of their affairs. Restrictions were also to be lifted upon the death of an allottee, and the law "gave the probate courts almost complete control of Indian administration."[36] Oklahoma, not insignificantly, also assumed the duty of protecting Indian minors.

The immediate result of the law of 1908 was to make more than 12 million acres of land subject to sale to whites and taxation by the state. It also re-established a system of leasing that had proven less than beneficial to the tribes in the past. This windfall came at the expense of the tribes. Instead of shifting protections from the federal government to the state, the law of 1908 facilitated dispossession. Many Indians whose restrictions were lifted after only cursory evaluations by officials did not possess much cash. As a result, many were forced to sell their lands to pay taxes. Others, who presumably still enjoyed a level of protection, were nonetheless victimized by land speculators and grafters who enjoyed a fairly unrestrained atmosphere in which to ply their trade. With editorials and politicians loudly advocating the end of Indian ownership and criticizing the "improper" use of land by ignorant savages, whites stepped up their assault on the tribes.[37]

Conservative members of the Five Tribes quailed in the face of these developments. They recognized that the 1908 law "meant paying taxes, potential confiscation of land because of bad debts, and the loss of government services."[38] In addition, the Oklahoma congressional delegation's victory appeared to signal the federal government's abdication of its responsibilities for Indian affairs. Writing on behalf of fellow Chickasaws and Choctaws in April 1910, Jim Frazier and Forbes D. Grouatt expressed alarm about the removal of restrictions from Indian lands. Although informed by their Congressman, Charles D. Carter, that they were quite capable to "manage our affairs and needed early settlement with the United States and Distribution of Tribal property and abolish the . . . Tribal School also," the two men maintained that "this is a wild mistake."[39]

Particularly troubling to these "fullblood petitioners," as Frazier and Grouatt identified themselves, was the lack of funds available to send their children to school. What they really desired, and fully expected, was that "the United States government be a guardian over our children from six years old to twenty one years old and educate in the work and school." Frazier and Grouatt wrote that "we are not prepared to handle our funds." Better that the government use those funds to pay for their children's' education. The petitioners maintained that "there are some competent person[s] that can take care are amongst our tribe," but these were few and they did not include most full bloods. They closed their appeal: "we have told you all what we know and seen it with our eyes[;] we are earnestly required you authority for some protection at you earliest convenient."[40]

Oklahoma officials proved deaf to such concerns. In fact, after their success in the wholesale lifting of restrictions in 1908, the state's congressmen in 1912 turned their attention to attacking the federal government itself. Oklahoma officials protested an appropriations bill that funded a network of district offices operated by the Indian bureau to better serve the populations of the Five Tribes. While their stated rationale was to consolidate the agency's operations and centralize administrative functions, the lawmakers' purpose was part of a broader effort to force the federal government to dismantle remaining protections on Indian land and other measures that kept Indians secure from state administrative functions.

Located throughout the state's eastern counties, the district offices brought the government closer to the tribal members. The offices, according to an article in the *Indian School Journal*, were part of an effort to place Indian bureau personnel near the people they served in order that they might come to better understand the needs of tribal members and deliver services more efficiently. Among other duties, the "District Agent must supervise the purchase of live stock, farming equipment, fencing material, etc., and see that good habitable houses are erected for them and that an exorbitant price is not charged for the same." The offices were also charged with evaluating applications for the removal of restrictions, "drawing up leases for the allottee and . . . investigating the adequacy of the consideration," handling complaints regarding lease arrangements, delivering payment to tribal members, "and encouraging the Indians to live on their land and farm the same."[41] In

Hardship and Decline 61

other words, the district offices worked to strengthen tribal members' hold on their land in direct opposition to the desires and efforts of whites.

The results of district office supervision were impressive. *The Indian School Journal* noted that "the District Agent is able to save thousands of dollars each year to the Indians, and not only this, but he is able to assist him in a way that will be of lasting benefit, not only to the Indian himself, but to the community in which he resides."[42] To prove the point, the article explained how the Madill, OK, office saved the local Indians 134,000 dollars over a four-year span. "[T]his saving was effected without wronging a single honest land purchaser or business man who deals with the Indian, and without a single dollar of cost to the Indian, either as an individual or tribe."[43]

The bureau's sixteen district offices were especially important in protecting the "minor and full-blood Indians of the Five Civilized Tribes." While noting that "[c]onsiderable doubt seems to exist in the minds of our white citizens as to the necessity" of the protections offered these "incompetent" Indians, the *Indian School Journal* argued that "too much emphasis cannot be placed on the need of the full-blood and minor Indians, and especially the minor who is just attaining his majority, for some means by which they can be instructed and warned against the snares of the grafter."[44]

Other sources, among them non-Indians, registered their support for the district offices and highlighted the dangers that existed for Indians if federal supervision was rescinded. The Vinita Retailers Association drafted resolutions to uphold the district offices. "[T]he machinery of the State is inadequate, and not organized so as to take the place of the protection heretofore afforded the class of Indians needing it."[45] The County Judge of Marshall County, J. W. Falkner, added to the chorus of protest that greeted the attempt to shut down the district offices. "I have had numerous occasions to call on the various District Indian Agents for their assistance in the matter of the estates of full blood Indians, both minors and adults, and I can truthfully state, that they have, almost without exception, rendered me valuable services in the discharge of my official duties pertaining to this class of citizens."[46] Falkner added that savings to the Indians were owed to the efforts of the district offices in "assisting them in the recovery of certain portions of their estates and also in enabling them to get fair and adequate considerations for the lands sold by them from inherited estates. . . . There are still many full bloods in this section of the state," the judge wrote, "who are absolutely incompetent to manage their property."[47] An equally large number of "designing and unscrupulous persons" were being foiled by the bureau. Falkner even admitted of the limitations of his own office in protecting Indians: "It is impossible for the County courts and other courts as well, to render full and complete protection to this class of citizens without the aid of some active Agency whose duty it is to make personal investigations of the individual cases."[48]

The sheriff of Hugo County wrote Sen. Robert Owen regarding the district offices and bluntly stated his protest. "Now Mr. Owens [sic] I am no grafter. I don't own one foot of farm lands in Oklahoma.... I have the interests of the Indians of this county, as well as the straight farmers and taxpayers, knowing the situation as well as I do, I feel that it is nothing but right for me to register a strenuous objection [to shutting down the district offices]."[49] A representative of the First National Bank in Bennington, OK., added that initially people were wary of the district offices. "In the beginning there was a good deal of prejudice in regard to the agents, it being the impression they were looking after the interests of the Indian regardless of the rights of anyone else, but that impression has been dispelled, and the relations between the two races has been much improved in a business way."[50] Clearly, the role occupied by the agency had found supporters in the general citizenry. The government performed duties that made relations more harmonious.

The district offices were, not surprisingly, well regarded by tribal members for providing much needed attention to their affairs. As the congressional battle heated up in 1912, the secretary of the Interior received a resolution from the Keetoowahs opposing the appropriation bill. "It is the unanimous sense of this Society," the document stated, "that all full blood Cherokees ... earnestly protest against the proposition.... It is the avowed duty of the United States to protect the full blood Indian, and the withdrawal at this time of the effective protection afforded them by the District Agents means not only the failure and refusal to protect, but in the opinion of this Society it will result in incalculable injury and loss to thousands of full blood Indians."[51] This type of Indian, Keetoowah leader Richard Wolfe telegrammed, "is incompetent to handle his affairs without the assistance of the District Agents, who have no interest but to do justice to the Indian. Without that help they are left to the mercy of the Speculators."[52]

The district office battle served as a lightning rod for discontent in other quarters as well. Conservatives discovered they were not alone among tribal members in decrying the attempt by white Oklahomans to dismantle yet another facet of federal protection. The general decline in the conditions of the tribe during the immediate post-statehood years sparked concern among progressive tribal members, many of whom were employed by the Indian bureau and were thus well informed of tribal affairs. In fact, after years of striving to distance themselves from full bloods, they began to try to shield them. "Through my veins courses Cherokee blood," wrote Cherokee national attorney W. W. Hastings in explaining his motivations for joining the district office fray. "I trust that I will not live long enough to see the full blood people of my race homeless in their own country, wanderers in their own land."[53] While economic changes in Oklahoma were viewed by progressives as formidable yet necessary challenges that demanded accommodation, white corruption and the aggressive tactics of state officials to gut federal protections for incompetent tribal members were galling. Progressives tended to view the federal government as an essential buffer between Indians and those who would dispossess them of their resources. It was the one institution that might preserve tribal members as they struggled to adapt to their changing

Hardship and Decline

conditions. "I am sure," added Hastings, "that no fair-minded citizen, white or red, whose good opinion is prized, will believe that these Indians should be set adrift, robbed of their lands without the assistance of the government."[54] "I am defending the system . . . and I am trying to arouse public sentiment against the wholesale attack on the system."[55]

The attack on the district offices provoked a spirited defense by progressive tribal members. In ending an editorial in the Ardmore Statesmen by stating that "[t]here are a few good fights in me yet, and I appeal to the public conscience for their assistance," M. L. Mott, a Creek who served as the tribe's attorney, wrote in August 1912 of his opposition to the proposed closure of the district offices.[56] "The Indians have no political influence," he wrote, "and the only protection they have is that afforded by the Government of the United States. The state [of Oklahoma] is powerless to protect them, however . . . honest and conscientious the officials may be in the discharge of their duties." To shut the district offices, Mott noted, "would be a great injustice to the Indians. The District Agents have saved to the Indians hundreds of thousands of dollars."[57]

Mott understood that the district offices were a principal means by which the government protected tribal members' land. "The Indians have given up everything practically, except their property rights. We have consented to the allotment of the land in severalty and have agreed by treaty to the abolishment of our Tribal Government. We have been made citizens of the United States and the State of Oklahoma." Yet despite these developments and their attendant promises of equality, Mott saw the link between Indians and the federal government as the most significant. The government, he argued, "provided for the appointment of District Agents to be stationed among the Indians of the Five Civilized Tribes for the purpose of protecting the property rights of the Indians."[58]

The Cherokee chief W. C. Rogers also argued for the retention of the district offices. Although he had been the target of conservative criticisms, as well as electoral challenges by the Keetoowahs, he fretted that conservative tribal members deserved greater protection, not less. In requesting that the efforts of the Oklahoma congressional delegation be stymied, Rogers wrote that "[i]t is the hope of the friends of this class of Indians that they will be dealt with patiently, and advised from time to time about their business affairs, in the hope that they will become much better able to care for their property interests."[59]

Although their efforts failed and the district offices ultimately succumbed to budgetary attacks, progressives continued to push for greater protections for indigent and otherwise suffering tribal members. They did so even as they continued to pursue upward mobility and individual success for themselves. Indeed, progressive activities on behalf of other tribal members squared with their own values and their support of the federal government as vital to Indian affairs. These apparent contradictions were perhaps most pointedly evident in the person of Gabe Parker, a Choctaw who in 1914 became Superintendent of the Five Tribes Agency.

Parker's was a notable appointment in that it signaled the boundless opportunities that appeared to exist for Indians in Oklahoma: aided by education, equipped with talent, and fortified by a willingness to work hard, an Indian could rise to prominence. Parker was the embodiment of the progressive ideal. Previously an educator and school supervisor, he was a product of the Dawes Era's philosophy of individualism, and he frequently wrote and spoke of the responsibility of Indians to pull themselves up by their bootstraps, to take advantage of the benefits of allotment. Tribalism, he believed, was an anachronism that had little place in modern life. Yet despite his devotion to assimilationist principles and practices, Parker labored tirelessly to use the Indian bureau for the betterment of his fellow tribesmen.

Gabe Parker worked mightily during the 1910s to implement federal programs to benefit the tribes. But he was paddling upriver. As he advanced policies to speed improvement, he was forced to battle institutions and individuals working to undermine his efforts. Even events seemingly conspired to foil his good works. The coming of World War I provided a number of opportunities to further limit federal protections for tribal members while enhancing state power. By placing urgent new priorities on the federal government, the war forced Indian affairs to the periphery where they became the province of state officials with relatively little federal oversight. For Parker the war, while appearing to provide Indians with the opportunity to improve themselves and their conditions, instead heightened the pressures on conservative tribal members.

Throughout the war, the process of awarding competency to Indians, and thus "freeing" them from governmental control, sped up. Consequently, more Indian land was subject to sale and lease. While Indians were eligible for competency following the 1906 passage of the Burke Act, an aggressive corollary called "forced competency" was introduced by Commissioner of Indian Affairs Robert Valentine, who served from 1909 to 1912. Frustrated by the Indians' reluctance to discard the cloak of federal protection, and pressured by increasingly powerful western politicians eager to rid their states of obstacles to development, Valentine "decided to force citizenship and fee simple titles on . . . Indians by a coercive application of the Burke Act." "Instead of waiting for allottees to apply for competency, he . . . form[ed] special commissions that would visit reservations, determine which individuals were capable of ending their trust status, and issue fee patents to the competent."[60]

Forced patenting was applied with disastrous results. First employed among the Omahas of Nebraska, it only hurried the process of separating Indians from their land, thereby benefiting a number of speculators patrolling the Omahas' reservation.[61] More importantly, the policy signaled a change in what competency meant and how it would be granted. Previously, the government desired to surround Indians with proper examples of civilized life and modes of behavior in order to introduce them to white norms. This would provide a way to uplift Indians by allowing them to appreciate the merits of coming into their individual affairs, of assimilating. To best define competency, a complex, often arduous process was instituted. But under forced competency, this time-consuming project was shortened. A competency certificate, it now appeared, was no longer regarded as something to be earned, something that was awarded because competency had been suitably demonstrated. Instead, it had become a piece of paper that was required in advance of actual competency, something that was needed in order to make future independence possible. An Indian would remain dependent, according to Valentine's design, unless he was forced to compete in the world. In this way, competency traveled a short but mean distance from being an achievement to being a pre-requisite for the lifting of essential protections.

Undeterred by the program's early failures, Valentine's successor, Cato Sells, an Iowa-born banker and Republican party figure in Texas, invigorated the policy in 1915. Sells was an energetic man who was responsive to demands by politicians to step up the Indian bureau's "individualization" campaign and to grant competency more liberally. He was informed in this attitude by a belief in the redemptive qualities of work. Reliance upon the government, in his view, led to moral degeneracy. Although he witnessed the ways that local land speculators perverted the intentions of federal policy by exploiting and dispossessing Indians, and the misery and hardship that resulted, Sells was unwavering in his dedication to the policies themselves. Toward those "whom he judged able to manage fully their affairs," Sells exhibited a "ruthless determination to rid the government of responsibility."[62] While he was also concerned for the Indians under his care—reasoning that "by eliminating Indians who did not need government care from the responsibility of the Indian Office, attention could then be concentrated on those wards who still needed help"[63]—he was more apt to limit the criteria required for competency, thus pushing more Indians into the mainstream.

Predictably, this policy exposed Indians to new troubles. By loosening restrictions, the Indian bureau invited new waves of speculators and grafters to descend on Indian allottees. In eastern Oklahoma, forced competency was implemented with devastating results. Already reeling from previous waves of dispossession, tribal members now had to contend with the government's avowed intention to reduce its defense of their interests. It soon became clear, even to many progressive tribal members who backed assimilationist principles, that the competency commission only heightened the dangers for most Indians.[64]

Gabe Parker initially welcomed the competency commission when it arrived in 1916. "In my opinion," he wrote, "nothing is more important than this work of the Competency Commission. If properly done it is a most important step in the right direction, if improperly done, it means a loss to really needy Indians of their only means of livelihood."[65] Unfortunately, in

Parker's view, the inexpert handling of the competency commission's duties by outsiders unfamiliar with the Five Tribes was leading to work "improperly done."[66]

As practiced by Special Indian agent O. M. McPherson, the head of the competency commission dispatched to eastern Oklahoma, the process of granting competency among the Five Tribes demonstrated few of the concerns that preoccupied earlier government officials, including Henry Dawes, the architect of allotment. Under McPherson's tenure, the implementation of competency tended to devolve upon statistical precision to the exclusion of other matters. Optimally, he wrote, the commission would remove restrictions "from 20 to 33 1/3 per cent of the restricted Indians" in any given place. Writing in 1917, McPherson reported that in several of the districts visited, he had managed to lift restrictions at an exacting pace: in Chickasha District "we recommended the removal of the restrictions of about 33 1/3 per cent;" and in the "Vinita District we have interviewed 190 restricted Indians; of this number we will recommend the removal of restrictions of 62, which is a fraction less than 33 per cent."[67] Competency had become little more than a mathematical exercise.

In response to McPherson's approach to the sensitive business of entrusting poorly-educated Indians with a certificate that would likely lead to the dispossession of what assets they still held, Parker spent a good deal of time trying to divine the consequences of forced competency. During 1917 and 1918 he culled reports from his field clerks and grew disillusioned with the competency commission. J. Marcus, who was employed by the Five Tribes agency but was detailed to work with the commission, indicated in October 1917 that the commissioners often "arbitrarily recommended" Indians for competency. In the course of interviewing some five hundred Cherokees, Marcus reported that while "[t]here were but very few differences of opinion among the Commission as to the competency of the various applicants . . . there was a wide difference of opinion of the two members of the Commission and myself with reference to making arbitrary recommendations, especially with reference to those who were of more than one half Cherokee Indian blood."[68]

Over time, Parker's sensitivity to the flaws in the commission's conduct only heightened and he questioned whether forced competency was suited to the ultimate objective of assimilating Indians. Using accounts submitted by bureau employees, he painted an ugly picture of what happened when restrictions were removed. "From time to time," he wrote, "reports have reached me concerning individual cases of Indians who should not have been removed from Departmental supervision." In one case, referring to a rule requiring newly christened competents to wait thirty days for final approval of their changed status, Parker wrote, "[i]t has been reported to me that a land buyer had eight or nine of these Indians at the Severs Hotel in Muskogee at one time, waiting for the thirty day period provided for in the orders for removal of restrictions to expire."[69] At the end of the period, the

Hardship and Decline

tribal members promptly signed away their holdings without discovering their potential worth on the open market.

To describe the effects of competency among Indians in the Sallisaw District, Parker forwarded reports from Thomas P. Roach, a bureau field clerk "who has been employed here for many years and who is a man of discretion." A typical sampling of Roach's findings included: "Mary Liver . . . sold her entire allotment of 160 acres to A. A. Taylor of Muskogee, for $240 and has nothing to show from the proceeds"; "Wilson Hatcher sold all of his land for a small consideration, made several trips to Joplin, Missouri, and returned with big loads of whiskey and has nothing to show from the sale of his land at this time"; "Sylvester Goingwolf sold 80 acres of his 90-acre allotment for $1000.00. He had been offered $1800.00 but he was taken to the woods by a prospective purchaser and was kept there until the 30 days expired when his land was bought for $1000.00. He does not read or write"; "Rabbit Coon sold 100 acres of his land to A. A. Taylor, of Muskogee for a very small consideration"; "Thomas Corntassel does not read, write or speak the English language. He sold his homestead for a small consideration. This was his only land"; "Smith Long does not speak, read or write English; was taken to Joplin, Missouri by F. A. Blank, of Stillwell, and sold his 90 acres for $500.00. When expenses were deducted about $400.00 remained and this has all been spent"; "Robin Dougherty sold his allotment for $7500.00. Oil Inspector reports the land to be worth between $9000.00 and $11000.00 for oil and gas purposes alone."[70]

The sum of these and other cases indicated to Parker that the stated goal of competency was not being achieved. In his correspondence, he stressed the importance of providing some means of survival for people unequipped to fend off the likes of unscrupulous land buyers. He wrote:

> The information obtained . . . shows that many of these Indians have disposed of the only property of value they owned. Because of lack of education and business experience they will never acquire any other property. Their allotments would have provided them and their families with a means of livelihood during their life but since they have disposed of their lands they are almost certain to become tenants on the lands of other people or day wage earners.[71]

The ability of speculators to sink their teeth into newly unrestricted Indians was well known to Parker and his employees. Not only did the names of several land dealers crop up time after time, but the means employed to induce Indians to sign over their freshly dated patents were becoming numbingly effective. The work of the competency commissions, it seems, was widely broadcast; there was no mystery as to when certain Indians were due to get their patents and the dealers flocked to them eagerly. Interest in Indian lands infected even the highest officeholders of the state. Oklahoma Governor Robert L. Williams, who supplemented his public duties with doses of land speculating, was among those kept abreast of the commission's work. In one instance, an underling sent him a letter and "a list of the Indian[s] that the Competency Board removed restrictions from. Thought it might be that you would be interested."[72] Undoubtedly Williams was, for he pursued various purchases of Indian allotments in 1917 and 1918.

Forced competency tested the faith of progressives in the federal government's commitment to Indians and its ability to prevent wide-spread depredation of Indian resources. They were further shaken by the developments accompanying World War I, when Indians confronted an infinitely larger tangle of problems with diminishing federal protection. The imperatives of national mobilization produced disorientation. Tribal members were subject to Selective Service and the draft, and even those who had not yet been granted citizenship as a result of allotment were frequently, and illegally, inducted.[73] Even as many progressives saw that the war held out the opportunity to promote Indian abilities and convince American society of Indian patriotism, it starkly revealed the insensitivity of government bureaucracies to Indian affairs.

Throughout the war, progressives strained to ensure a degree of special treatment for conservative tribal members. Gabe Parker, handicapped by shrinking numbers of bureau employees at his disposal, labored to inform tribal members of their rights for exemption from the draft.[74] While he worked in earnest to get Indian men to register with Selective Service and volunteer for service, he also personally dispatched investigators to guarantee that Indians received fair hearings on requests for hardship leave. He frequently wrote his superiors and military officials to gain benefits for tribal members. In one case, Parker sent a special bureau agent to Texas to provide advice to a Choctaw man convicted in a court martial proceeding of murdering a French farmer while serving abroad with the American Expeditionary Force. Parker expressed concern that non-Indian military personnel with little understanding of Indians and Indian behavior might have misinterpreted the Choctaw's stoic and silent acceptance of the criminal charges as evidence of guilt rather than a manifestation of cultural background.[75] Another influential progressive, G. W. Grayson, principal chief of the Creeks, petitioned the War Department to grant exemptions to full blood Creeks as they had an aversion to crossing the ocean to fight in Europe.[76] For Grayson, Parker, and a host of others, the demands of the war did not overshadow the obligations they felt toward their fellow tribesmen. Although outwardly supportive of assimilationist policies, they often tried to subvert government plans, as well as the activities of white grafters and Oklahoma officials, by arguing that some tribal members needed more time to become conditioned to life in modern society. Throughout World War I, they remained on the alert for any instances of discrimination and quickly maneuvered to gain tribal members a just hearing in cases of trouble.[77]

* * *

Hardship and Decline

During the post-statehood period, progressive tribal members grew disillusioned with the promises of the Dawes reforms and the dedication of federal officials to carry them out. But despite the hardships of many tribal members, a fuller identification by progressives with their fellow tribesmen remained stillborn. The injustices suffered by conservative tribal members remained a somewhat abstract notion to many progressives in the early post-statehood period. While they required protection because they did not possess the wherewithal to compete, in their refusal to accept allotment, to attend formal schools, and to embrace new economic opportunities, conservatives invited a measure of hardship upon themselves. The defiant rhetoric of separatism yielded persecution. Progressives, in contrast, had fought hard in the first years of the twentieth century to escape the burdens placed upon Indians in American society.

Yet there existed an obligation to assist conservatives in their troubles. Men like Gabe Parker considered it necessary to work for the betterment of those who lacked the power to help themselves. Progressives' own accomplishments seemed inescapably tarnished when weighed against the overwhelming disaster visited upon many of their fellow tribesmen. For W. W. Hastings, the Cherokee national attorney, "the best years of my life have been given to the Cherokee people," especially "the admittedly incompetent fullblood Indian." One of the more powerful explanations of his devotion, it appeared, resided in his simple statement that "[t]hrough my veins courses Cherokee blood." Although progressives still pointed to and perpetuated distinctions among tribal members, and clung to their whiteness in contrast to the "real" Indians' Indianness, the gulf between tribal members had narrowed.

SIX
Assimilation's Failure

Victor Locke, Jr., peeled off his military uniform, unfastened the clusters signifying the rank of major, put aside his visored officer's cap, and returned home from war intending pick up where he had left off. The son of a white man and Choctaw woman, Locke exemplified the successful, progressive Indian. He was equipped with a network of important contacts, a college education, and experiences and skills that allowed him to navigate both the white and Choctaw worlds. A former Oklahoma state legislator and principal chief of the Choctaws, the owner of a distinguished military record, having achieved the highest rank of any Indian in the army, he seemed destined for a notable political career. Immediately after the war, Locke's name possessed rare currency and was bandied about in discussions for the post of Commissioner for Indian Affairs. Although he did not receive that appointment, he was named superintendent for the Five Civilized Tribes in June 1921.[1]

Locke, however, did not go on to blaze a path in the Indian service. Within two years of taking over the bureau's offices in Muskogee, he found himself out of a job. Hardly the first politician to run afoul of party patronage expectations, Locke undermined his own superintendency after discovering the extent to which white Oklahomans had plundered tribal assets. His crime, one writer acidly noted, was to respond to these practices by adopting "a policy devoted to the interest of the Indians." The Choctaw war hero "attempted to administer the Agency with an independence that soon brought him into disfavor with the spoilsmen."[2] His efforts to clean up the corrupt handling of Indian resources provoked angry responses. When he blocked several men from becoming probate attorneys, his career with the Indian bureau was quickly terminated.[3]

Locke was hardly a naive victim of "underground forces."[4] Rather, he mindfully challenged the conventions of Indian affairs. Unlike his predecessor, Gabe Parker, who worked cautiously within the framework of his authority as an employee of the Indian bureau, Locke challenged such limitations. He did so because the conditions in which tribal members lived continued to greatly deteriorate through World War I and into the 1920s, a development that discomfitingly encroached not only on conservatives but upon progressives as well. Progressives grew alarmed that they, too, were being denied the upward mobility they had anticipated just a few years earlier.

Tribal members found the path to acceptance and prosperity blocked. The federal government retreated further from its responsibilities to protect Indians and their resources. Social conformity, initiated by the patriotic fervor of World War I, grew virulent in the 1920s until a stifling intolerance bred discrimination and oppression. Worse, as the state's economy matured, concentrating upon mineral extraction and large-scale farming, many tribal members were targeted for their oil holdings. Oil, it became clear, provided an even greater lure for grafters than farm land; Oklahoma chased oil rights in a frenzy of exploitation and violence that made their previous activities among the tribes seem almost friendly. Sensing that restraint and the will to work within the system had failed, progressives began to venture beyond the usual protocols for administering federal Indian policy.

The developments of the immediate post-war period affected tribal relations in a novel way. As white Oklahomans increasingly ignored the distinctions between tribal members and the government grew less patient with the slow conversion of Indians into American citizens, diverse segments of the Five Tribes populations began to address their internal divisions. Recognizing the limitations of their previous strategies in dealing with white society, progressives and conservatives alike reached out, tentatively at first, to one another. This effort would require that tribal members reconcile their differences and disagreements, including those of blood, and work toward solidarity.

* * *

If the trends of tribal decline appeared distant to the concerns of many progressives in the years prior to World War I, by the 1920s they had drawn noticeably closer. Indeed, progressives were personally affected by several developments, not the least of which was the dispiriting performance of the federal government in safeguarding tribal interests. Progressives were achingly aware that the surest defender of tribal interests was in the midst of withdrawing from the field. This was true throughout the country, where the increasing power of western politicians was directed toward assisting non-Indian constituents at the expense of Indians. The selection of Albert Fall, a westerner from New Mexico, to lead the Department of the Interior in the

1920s was but a single example of the emerging influence of officials from the country's newest region. Fall engaged in an active campaign to open up formerly closed lands for development, including for oil and gas exploration and leasing. With the government standing behind those who wished to exploit Indian resources, the debilitating trends in Indian policy continued.[5]

Such policies were backed by a shift in the intellectual assessment of Indians and their role in American society. During the late nineteenth century, federal policy was marked to a strong degree by hopefulness in the ability of Indians to be transformed into productive citizens. Although the execution of government plans was often marred by the venal actions of Indian bureau personnel, optimism nonetheless fired the imaginations of earlier reformers and activists. They worked for programs that had as their objective the uplifting of Indians from savagery to civilization; they embraced the possibility of assimilation, of bringing Indians into the melting pot of American society. The general ideas informing this viewpoint and which gave it credibility stemmed from the belief that Indians were capable of a relatively swift conversion. Through education, Christianity, and individual landownership Indians could gain equality within mainstream society. They were, in rough terms, suitable raw material for this ambitious plan. This philosophy was the hallmark of former Indian formulas such as the post-Civil War Peace Policy of President Ulysses S. Grant and its successor, allotment. Both of these programs relied for their success on the Indian's potential for reform and improvement.[6]

But increasingly by the 1920s, the perfectionist sensibilities of old-line reformers such as Henry Dawes were viewed as quixotic. The goal of assimilation, under fire by Indians who felt betrayed by the unwillingness of whites to fulfill their promises, also drew criticism from government officials who suspected that Indians were not the proper material for such an experiment. The realization dawned that assimilation could not be achieved through a simple substitution of clean clothes for blankets or bibles for ceremonial trinkets. These exchanges did not guarantee racial improvement. While there were exceptions, and some Indians were conspicuous in their success, they only proved the rule. As a result, well-intentioned policies, dependent upon the notion that Indians had the potential to become Americans, were revised and downgraded; Indians simply could not be rushed into the modern era.[7]

Among Indian bureau officials, a "pessimistic" outlook toward Indians gained sway and commitment to former principles flagged. More and more, policy makers saw the "Indian problem" as intractable. Despairing at the slow pace at which Indians seemed to be adjusting to civilized life, there was a sense that perhaps Indians could only be expected to progress so far; perhaps they could be domesticated and put to work, but grand plans for their equal partnership in American society needed to be shelved. Even as they clung to the old methods of reformers, including an insistence on individual landownership, haircuts, and a dose of religion, officials began to hasten the timetable for Indian conversion. The government, it appeared, was engaged in an end-game strategy. One line of thought among Indian bureau officials was that the government should aim to get out of the Indian business, to wrap up its affairs and abandon a decisively unfruitful project. To this end, the Indian bureau stepped up its effort to usher Indians into the mainstream while lifting its protections on Indian lands and interests. The competency campaign, and other programs, was evidence of this urgency.[8]

In eastern Oklahoma, despite attempts by the Indian bureau to inhibit the most egregious abuses such as those perpetrated by the probate courts, the government's influence waned. In part, this resulted from an increasingly aggressive and confident state government. The tide in the struggle for control over Indian resources began to turn shortly after the war, when, in 1920, the Republicans swept into office and "upset the precarious balance that had been established between the state and federal governments."[9] "Because of the sordid influences controlling appointments to the Indian service the protection offered by Federal Government probably reached its lowest level during this period."[10]

For progressives, the federal retreat manifested itself in, among other areas, education. In earlier eras, progressive tribal members built and ran tribal schools. Even after allotment and statehood, when the government took over Indian schools and public schools opened their doors to Indians, they encouraged their children to excel. In turn, progressives expected excellence in the facilities and instruction their children received. They expected rigorous, academic curricula of the kind they had emphasized in Indian Territory.[11] As a child, for example, the Creek writer Alex Posey recalled studying Byron and Keats and other leading figures of classical Western education.[12]

But by the 1920s, as other barometers of Indian life dipped, progressives saw their children's educational prospects darken. Government-run Indian schools, following a trajectory similar to that of other federal Indian programs, began to institute more vocational training at the expense of academic subjects. One analyst of federal school policy has observed:

> During the early twentieth century new political, scientific, and social attitudes created a climate of opinion that shattered the previously widespread faith in the schoolroom's ability to solve the Indian Problem. . . . Specifically critics questioned the suitability of educating Indians if such schooling were inconsistent with the more narrowly defined roles that social scientists and policy experts assigned to Indians.[13]

In his study of the Chilocco Indian school in north-central Oklahoma, a school with large numbers of children from the Five Tribes, K. Tsianina Lomawaima notes that students and their parents in the 1910s and 1920s became aware that they were being groomed not for the professions but for a second-class citizenship of drudgery and manual labor. As Lomawaima maintains, "federal boarding schools did not train Indian youth for assimila-

Assimilation's Failure

tion into the American melting pot, but trained them in the work discipline of the Protestant ethic, to accept their proper place in society as a marginal class."[14] This occurred despite the rising numbers of mixed blood students at Chilocco, that portion of the population that was expected to perform better in the classroom than their full blood counterparts.[15]

Beyond deteriorating opportunities in education, other warning signs proliferated. Progressives were attuned to a rising tide of intolerance in Oklahoma. Part of a broader strain of racism in Oklahoma that intensified during World War I, anti-Indian attitudes in the 1920s threatened to obliterate the distinctions between tribal members. Fueled by an increasing resentment of Indians by whites, a resentment that increased further as oil wells began to bring great wealth to a handful of tribal allottees, this intolerance swept all Indians together into a single racial category.[16]

During World War I, Oklahomans fostered a climate of oppression.[17] Whipped into a patriotic lather, they squelched, often violently, the faintest hint of dissent. Loyalty campaigns, spawned by notions of "100 percent Americanism," erupted throughout the state during the war. In overt bids to contain disorder, these campaigns targeted German-Americans, socialists, labor unionists, and others suspected of disseminating traitorous ideas or ideologies.[18] For Indians, the barest suggestion of dissent was dangerous. In one well publicized instance, a number of tribal members joined with a radical tenant farmer organization in protest of military conscription and of the "capitalistic" causes of the war in Europe. In what came to be known as the Green Corn Rebellion, small groups of farmers belonging to the Working Class Union took up arms and planned to gather together in a larger movement before marching to Washington to register their complaints. They would eat the green corn not yet ready for harvesting for sustenance. Although the rebellion nearly ended before it began, and local authorities quickly moved to end the trouble, Indian participation greatly disturbed the Indian bureau and other federal officials. Warrants were issued for the alleged tribal ring leaders and federal agencies went to significant lengths to determine the extent of Indian involvement. The message to tribal members was clear: resistance prompted backlash and was ultimately futile.[19]

What began as a campaign to ensure war-time loyalty, however, continued in Oklahoma after the war. Efforts to promote a rigid social conformity spawned a variety of movements and organizations and created an oppressive environment.[20] By the early 1920s, the state was home to one of the nation's strongest, and most violent, branches of the Ku Klux Klan. The Klan ran a powerful political machine and occupied numerous state and county offices. In various locales, its members included leading figures in government and business. According to scholars, the Klan operated as an enforcer of social codes and the keeper of morality and decency. In so doing, it was often engaged in carrying out retributive measures against such social nuisances as wife beaters and drunks. The Klan also limited the unseemly influences of cosmopolitan America by banning certain kinds of movies in local theaters and speaking out against authors of "subversive" and "pornographic" books.[21]

But of course the Klan also reflected intensifying racial tensions in the state. It had not changed so much from its Reconstruction Era predecessor that it did not exploit anti-black sentiments and negative attitudes directed against other minority groups. When, in 1921, a black man in Tulsa, the state's most dynamic city and home of the burgeoning oil industry, was arrested for attacking a white man, the Klan and other organizations were in the front lines of enraged white citizens. The result was a vicious race riot that claimed the lives of several blacks and led to the burning of a large portion of the black section of town.[22]

The rising tide of intolerance also swamped Indians. This was perhaps best illustrated in the developments accompanying an oil boom that occurred in the waning years of World War I and accelerated through the 1920s. The liquid gold that bubbled out of the ground spurred a venomous hatred of many whites toward their Indian neighbors. Whereas resentment of many white migrants in the late nineteenth century toward their Indian landlords was kept under control by the sovereignty of the tribal nations, by the 1920s, resentment among whites for those Indians sitting atop valuable allotments could not be contained.

The oil industry had grown steadily in the years immediately preceding World War I, but with armies mobilizing and heading into battle, the need for heavy equipment and fuel sources rose rapidly. And demand did not relent with the advent of peace. While agriculture and other economic sectors suffered a setback in the early 1920s, oil production reached dizzying heights. With the revolutionary re-organization of factory work and the introduction of assembly line-manufacture of automobiles, trucks, and motorized agricultural equipment, a confluence of factors transformed oil into a potent source of strike-it-rich opportunity. With a large market beckoning, interest in the profitability of the industry soon attracted entrepreneurs and large corporations to Oklahoma in a race for gushers.[23] Boomtowns popped up overnight throughout the state. These ramshackle towns held men with fantastic dreams of wealth and fostered behavior last seen in the gold camps of California and Alaska. The surging numbers of oil workers also testified to the changing shape of Oklahoma's economy and the allure of wage work both for men choosing to seek their fortune and those forced off their farms by slumping agricultural prices.

Eastern Oklahoma boasted an abundance of oil, and several substantial underground pools were first discovered in the 1850s. The first commercial field began production in 1896 at Bartlesville in Osage County.[24] Additional fields came on line soon after at Tulsa and areas surrounding Bartlesville. By 1905 the output from twenty five fields exceeded eight million barrels. By 1915 the flow of crude from more than one hundred fields topped one hundred and eighteen million barrels, and Oklahoma soon surpassed the billion-barrel mark.[25] The discovery of several "giant oil fields" placed the state in the

Assimilation's Failure

top rank of petroleum-producers. The Glenn, Cushing, Healdton, Sho-Vel-Tum, Burbank, and Seminole Pools were all discovered between 1905 and 1923.[26] From 1906 to 1928, Oklahoma was the leading oil producer among southwestern states, contributing roughly 30 percent of all of America's oil. By 1929, its total mineral production, including coal, was second only to Pennsylvania, and this segment of the economy contributed nearly 15 percent of all Oklahoma incomes by the late 1920s.[27]

Initially, it appeared that the oil strikes, many of which were located on Indian-owned land, might benefit various tribes in Oklahoma. Among the Osages, for example, the Burbank and Avant fields promised to bring untold wealth. Near neighbors of the Cherokees and Creeks on the Kansas border, the Osages by the 1920s were recognized as the richest group of people, per capita, in the world. They were the accidental benefactors of events taking place elsewhere in the nation that stimulated a growing demand for oil. In turn, those distant developments soon transformed the capitol of Osage County, Bartlesville, and the Osages beyond recognition. From a sleepy county seat, Bartlesville emerged in the early 1920s as an oasis of fabulous wealth. Its streets were lined with stores boasting the latest fashions from Paris and New York. Expensive automobiles cruised the town shuttling Indian "princesses" from boutique fittings to freshly built mansions to late night parties in luxurious hotels. Banks and corporate headquarters, equipped with the latest technologies and appointed with the finest marble, were monuments to the vast "underground reservoir" of liquid riches.[28]

But there was another, seemier side to the oil rush. Just as the boomtowns displayed tendencies toward violence and loose morality, the oil industry seemed to bring out the worst in human nature. As other mineral rushes had done in the past, it sparked greed and spawned misery. For every man who struck it rich, thousands achieved far less glamorous results.[29]

Yet, without doubt, the industry's heaviest toll fell upon Oklahoma's Indians. Alongside the grandeur of luxury hotels and Rolls Royce automobiles was a mirror world of crime and corruption. Owners of Osage "headrights," individual rights to the tribe's commonly held mineral resources, became the targets of innumerable criminal acts ranging from the petty to the homicidal. Local salesmen, for instance, often raised prices when selling to wealthy Osages, and attorneys' fees for even the smallest tidbit of legal work were extravagant. More serious were the actions of Osage guardians, those whites who were appointed by the courts to manage the interests of minors and "incompetent" wards. More often than not, guardians converted their duty to care for their wards into lucrative propositions whereby they looted property and money. In addition, men often traveled to Oklahoma strictly to marry Osage wives who were endowed with particularly valuable headrights.[30]

The Osages also fell victim to more violent crimes. In the early 1920s, more than sixty men, women, and children were killed, often by non-Indian members of their family, in a spree that ultimately brought federal criminal investigators to Oklahoma to conduct what came to be known as the "Reign of Terror" investigation.[31] Not only, it turned out, were Osage women defrauded by their white husbands, but a number were killed by them. In addition, some of the children "adopted" by white guardians lost more than their inheritance. There were cases in which the children simply disappeared never to be heard from again.

The tales of murder and corruption were outlandish, in part because of the fact that few, if any, of the suspects in these crimes were ever convicted. Protections were scant for Indians, and the murders among the Osages often went uninvestigated or unprosecuted. They were not hushed up, as there was little need to hide what everyone suspected, but were instead acts brazenly committed. Many Indians, petrified by their vulnerability and the violence it inspired, fled the state. Despite the eventual arrival of the Federal Bureau of Investigation, few perpetrators were held accountable for their deeds. In fact, whites acted without fear of imprisonment. Some offered testimony in court, testimony potentially damaging to themselves, that was surprisingly graphic and honest. They told matter-of-factly how they killed and why. In one courtroom scene, Kelsie Morrison, a "youthful convict" on trial for his role in helping to kill the heir to a large oil estate, related the planning of the murder. After getting his victim drunk on whiskey, Morrison told the court, he held her arms behind her back while his associate shot her in the head. One newspaper reported that the crime, carried out in a ravine outside of Pawhuska, Ok., was conceived and carried out as a "coolly calculated business proposition."[32] Neither man was convicted.

What became abundantly clear in the Osage Nation was the resentment that whites bore toward Indians. Oklahoma whites could not hide their frustration when presented with examples of "unearned" Indian wealth. "Backward" Indians had no legitimate right to lands they could not develop to their fullest potential. Or so the argument went. Indians were regarded merely as obstacles to white settlement and as beneficiaries of misguided government policy. One letter writer to Bartlesville's *The Independent* newspaper noted in 1926 the popular attitude toward wealthy Indians.

> It is easy for any Nordic down here [in Tulsa] to understand the outraged feelings with which a poor Oklahoman, perhaps belonging to the Ku Klux Klan, looks upon the Osage in his mansion, surrounded by parked motor cars in bright colors. 'There,' says the Nordic, 'is a good-for-nothing redskin. . . . This wastrel is rich through no virtue of his own, but merely because the Government unfortunately located him upon oil land which we white folks have developed for him.'[33]

While the writer maintained that he did not sanction the violence visited upon the Osages, "on the other hand, I can perceive no boon, either of utility or beauty, in preserving them inviolate, like so many hothouse violets, from the competitions the rest of us must face in body and estate."[34]

Assimilation's Failure

According to this way of thinking, crimes against Indians were blamed on the Indians' own ignorance and gullibility. Because Indians were naturally lazy and thriftless, and because they did not deserve their good fortune, the fact that they were the subjects of fraudulent schemes and physical violence was, in part, their own fault. It was destined that "savages" would wind up with nothing since they did not understand the value either of money or honest work. Such was the self-fulfilling prophecy of many white observers. The lives of Indians simply mattered less than those of whites in Oklahoma. Moreover, attempts by the federal government to engineer race-based policies offended whites. Many whites considered themselves caught between an elite class on the one hand and a racial group protected by a distant and unknowledgeable federal government on the other.[35]

For members of the Five Tribes, who sat atop some of the largest oil fields in the state, the vast bounty lying undisturbed beneath their hardscrabble allotments proved an unsuspected source of malevolence. What might have been an incalculable material asset became instead a fearsome liability. Not only did Indians fail to reap a just share of the mineral wealth, but the very richness of their resources invited more intense persecution. The Five Tribes' oil drew to them the ugliest elements of corruption and greed. While their tale is somewhat different from that of the Osages, in that they did not collectively own mineral rights below ground but rather possessed individual claims to allotments, it contained many of the same elements. Tribal allotments attracted fierce contests among speculators and incited behavior and attitudes toward Indians that were as detrimental to the Five Tribes as those that overwhelmed the Osages.

Land speculation and mineral leasing became a high stakes gamble. Tribal allotments drew the attention of men who frequently skirted the law in the pell mell rush to obtain rights to potentially enormous oil reserves. Consequently, members of the Five Tribes were reckoned as a hindrance to be dealt with as summarily as possible, a minor part of the price of doing business in the oil industry. They were to be courted, seduced, and influenced to sell or lease, generally at extremely disadvantaged terms. If they proved stubborn, they were to be coerced. Coercion took a variety of forms, including getting Indian allottees drunk and forging signatures on land titles. Other measures called for men, women, or children to be kidnapped, ferried across state lines, and held until they agreed to sell.[36]

The atrocities of oil speculation against members of the Five Tribes knew few bounds. In 1929, four Choctaw women died under circumstances which led officials "to suspect foul play, probably poisoning." Although it was not ultimately proven that the women, whose land was coveted by "professional land buyers," were poisoned, the tactic was sufficiently well known in the region to be considered a possible cause of death in suspicious cases.[37] Landowning Indians not infrequently turned up dead with empty bottles of whiskey or moonshine nearby. While the presence of alcohol provided investigators with evidence for a quick dismissal of the case—a drunk Indian exposed to the elements provided little cause for greater scrutiny—it was routinely conceded, even boasted, that the moonshine had been doctored with wood alcohol or other lethal compounds. Considering the widespread problems of alcoholism among the Indian and white populations, this trick was a safe method of killing off Indians, for autopsies would not be held to investigate apparent drunkenness.

In his report on the deaths of the four Choctaw women, an Indian bureau field clerk wrote that the counties of the Choctaw nation "are overrun with men who are in the market for inherited allotments." These men staked out local hotels when elderly allottees were reportedly nearing their deaths. "In fact," the clerk wrote, "the methods of the men whose business is to buy up land as soon as death makes it available, remind one of a flock of buzzards soaring above the body of a dying calf." And while the clerk concluded that the deaths in this instance were likely not the result of poisoning, "it is unfortunately true that at least one home was dynamited for the purpose of land grabbing. . . . It is by no means unbelievable that in the Choctaw country, so far as inherited Indian lands are to be considered, avarice is securely enthroned and the rights of Indians, especially helpless children, are in danger."[38]

Indians had abundant reason to feel insecure in eastern Oklahoma. In the face of the rising death toll accompanying grafting practices, the federal government as much as admitted its powerlessness to halt the slaughter. "The Indian Service will unquestionably suffer severe criticism," the field clerk wrote in reference to the Choctaw women's deaths, "but under existing law the Service, although fully informed of what is going on, is powerless to do more than discourage the methods in vogue and place as many obstacles in the path of the land grabber as possible."[39] Tribal members, regardless of their background, came to understand that they received fewer protections under the law than other citizens. In 1919, for example, Burney Underwood, a Chickasaw, was beaten and shot to death by a policeman as he lay inebriated on a railroad station platform. Underwood had earlier in the day cashed a check for $150 and had gone on a drinking binge. The policeman beat Underwood with his "sixshooter," and then, according to one Indian bureau report, "threatened to shoot Underwood if he did not get up and walk and then deliberately shot him, Underwood dying about twelve hours thereafter." Witnesses said that "one of the blows on the head would have caused the death of Underwood had he not been shot, and that Underwood was apparently . . . helpless and the killing was unprovoked."[40] Law enforcement was suspect at best, and Indians understood their vulnerability before local authority.

In 1925, a report submitted by the Board of Indian Commissioners, which was investigating claims of corruption, noted that "[g]rafting on rich Indians has become a recognized business in eastern Oklahoma, and a considerable class of unscrupulous individuals live by this method. So common is this practice, that the term 'grafter' carries little opprobrium with it in

Assimilation's Failure

Oklahoma, and is the commonly used designation for those who have dealings with the Indians."[41]

In the previous year, 1924, the Indian Rights Association, a philanthropic organization based in Philadelphia, also investigated the probate system and it offered damning evidence and righteous accusations. In the association's report, entitled "Oklahoma's Poor Rich Indians," Matthew K. Sniffen, secretary of the association, Gertrude Bonnin, a Sioux who was listed as a Research Agent with the Indian Welfare Committee of the General Federation of Women's Clubs, and Charles H. Fabens, of the American Indian Defense Association, railed against the corruption of the probate courts of eastern Oklahoma. Subtitled "An Orgy of Graft and Exploitation of the Five Civilized Tribes—Legalized Robbery," the 39-page tract reported findings from a five-week tour of the region taken in 1923 as well as offering statistics on competency and landholdings and a capsule history of the mistreatment of tribal members.[42]

Since 1908, according to the investigation, the tribes had been "shamelessly and openly robbed in a scientific and ruthless manner." The costs of administering Indian assets were absurdly large and constituted an ill-disguised system of extortion. This corrupt machine was deftly operated by interweaving networks of "county judges, guardians, attorneys, bankers, merchants—not even overlooking the undertaker." Indians and their resources were treated as "legitimate game," as little more than a source of wealth to be tapped and administered for the profit of the individuals and institutions in charge.[43]

The tone of disgust in which the report was written stemmed from the astounding discoveries of the researchers. Greed-inspired guardians had failed to take even the most basic measures to ensure the welfare of their charges. "Indian children," the pamphlet recorded, "have been allowed to die for the lack of nourishment because of the heartlessness and indifference of their professional guardians, who had ample funds in their possession for the care of the wards." One attorney, it was reported, "received $35,000 from a ward's estate, and never appeared in court." The discovery of oil on an Indian's land "is usually considered prima facie evidence that he is incompetent, and in the appointment of a guardian for him his wishes in the matter are rarely considered."[44]

Venality marked the behavior of "professional" guardians. "[Y]oung Indian girls (mere children in size and mentality) have been robbed of their virtue and their property through kidnapping and a liberal use of liquor."[45] In several cases it was revealed how guardians and grafters went about their work. Millie Neharkey, of Tulsa County, described as an "18 year old Indian girl," was "kidnapped a few days prior to her reaching legal age." The owner of lands valued at $150,000, Neharkey was spirited away to Missouri by representatives of the Gladys Belle Oil Company and induced to sign papers that would give them power of attorney. After plying the girl with liquor as the party traveled through the Ozarks before returning to Oklahoma, the kidnappers gained title to the girl's lands and, according to some, may have assaulted her as well. Gertrude Bonnin, who met with Neharkey, wrote that she "grew dumb at the horrible things she rehearsed. . . . There was nothing I could say. Mutely I put my arms around her, whose great wealth had made her a victim of an unscrupulous, lawless party, and whose little body was mutilated by a drunken fiend who assaulted her night after night."[46]

The extent to which guardians obtained virtual control over their wards' lives was stunning. In recounting the story of Ledcie Stechi, a seven-year-old Choctaw girl who inherited oil-rich lands from her mother, Bonnin described, perhaps exaggeratingly, how the girl and her grandmother were kept at a level of near-starvation. After a portion of her allotment was sold for about $2,000, the McCurtain County judge limited Stechi to $10 a month for support. When a guardian was appointed to look after the girl's increasingly productive acreage, $15 a month was allowed in credit at a local store. Stechi's grandmother had to walk to the store, which was situated some distance from the girl's home in the hills, and was often too tired to walk home as the guardian made no allowance for transportation. "During [the] period from 1921 to 1923, the guardian did nothing to make them more comfortable or to educate the little Indian heiress."[47]

When Stechi's allotment was appraised at $90,000, the girl's allowance was increased to $200 a month, but whether or not that money made it to the girl and her grandmother was doubtful. When she arrived in Idabel, the County seat, in April 1923 to attend school, she offered a heart-breaking spectacle. "The little rich Choctaw girl, with her feeble grandmother, came to town carrying their clothes, a bundle of faded rags, in a flour sack. Ledcie was dirty, filthy, and covered with vermin. She was emaciated and weighed about 47 pounds." A medical examination given by the Indian service revealed, unsurprisingly, that Stechi was malnourished and also "poisoned by malaria." After five weeks of close medical supervision, the girl was placed in an Indian school, whereupon her guardian panicked. Fearing that he might lose control of his ward, the guardian demanded Stechi be delivered to his care, a request that was granted. A month later . . . word was brought to the hills that Ledcie was dead. There had been no word of the girl's illness and the sudden news of her death was a terrible blow to the poor old grandmother."[48]

When the corpse was brought to the grandmother's house, it was accompanied by a pack of grafters who hoped to gain what valuable property remained. "Greed for the girl's lands and oil property actuated the grafters and made them like beasts surrounding their prey," Bonnin's report sensationally recounted. When the grandmother requested an autopsy of the body, for she suspected that her granddaughter had been poisoned, she was rebuffed. "Feebly, hopelessly, she wailed over the little dead body—its baby mouth turned black, little fingernails turned black, and even the little breast turned black!" The grafters continued to harass the grandmother as did the courts, which appointed a guardian to look after her affairs, as she was

Assimilation's Failure

deemed incompetent. Her "vehement protest" against guardian supervision fell on deaf ears.[49]

Minors were not the only Indians targeted. According to one investigation in the 1920s, "to the grafters 'the quick and the dead' are all the same. On one occasion they waited literally at the bedside of a dying woman, and hardly had the breath left her body when her thumb was pressed upon an ink pad and an impression from it made on an alleged will, which was later offered for probate."[50] The threat to the material interests of tribal members, and the orderly system of inheritance by which families handed down wealth to their children and loved ones, was overwhelming.

Another investigation, launched at about the same time that Sniffen, Bonnin, and Fabens were touring eastern Oklahoma, was headed up by Edward Merrick, who was in charge of the law division of the Indian bureau's Muskogee office. Although less sensational in its findings, it nevertheless uncovered corruption in the system by which whites manipulated Indians. He examined the disposition of full blood inheritance sales by the county courts for the decade from 1912 to 1922 and compiled a variety of telling statistics. By tracking down and calculating the loans secured on property that was sold by Indians to non-Indians, Merrick demonstrated that the value of the loans often far exceeded the price paid to the Indian seller. In addition, he showed that administration costs pocketed by guardians were often upwards of fifty percent of the value of a given Indian estate.[51]

Even in cases where Indians were not harmed and were able to keep large portions of their wealth, they seemed to be victimized. While violence and fraud were the methods of choice exercised by some grafters, there were more subtle methods of stripping wealth from tribal members. Indeed, even in the absence of crooked guardians, the very nature of the law and local courts seemed geared against the interests of Indians. The legal infrastructure appeared a complicated mechanism designed to fleece the poorly educated and unwary individual or to overwhelm an otherwise competent Indian. The courts at times hosted a circus-like environment in which oil-rich Indians were set upon by any number of claimants desiring a share of the profits. The most famous case involved a Creek named Jackson Barnett, whose allotment was the site of one of the greatest oil strikes in the region. A man with minimal education and a bare grasp of English, Barnett soon found himself at the center of incomprehensible legal wranglings. Everyone, it seemed, clamored for a piece of Barnett's pie. The government, numerous sharpers, a young non-Indian wife who swept him off to southern California, an ever-increasing number of alleged relatives and heirs, and a host of others amassed legal briefs and motions by the yard. Every time Barnett wanted to spend money, build a house, give to charity, or travel across state lines he was hounded by red tape.[52]

As the instances of "legal criminality" suggest, much of the looting of tribal resources in the 1920s was the work of organized systems of relations. At base, the plunder was enabled by structures established by whites to serve their interests while precluding those of Indians. As the profitability of oil rose, many speculators, oil companies, and local officials tried to forestall their risks in dealing with tribal members. In an attempt to rationalize their ventures, the assembled interests rigged a system that effectively reduced their chances of failure. In eastern Oklahoma, while individual acts of violence continued to be used against tribal members, a more subtle form of dispossession emerged that was based upon patronage and corruption. In a complex web of relations, county courts and local officials were tied to oil companies, attorneys, and speculators. Judges were bribed to authorize shady land sales or lease agreements. Similarly, as in earlier years, judges routinely awarded guardianships over Indian minors to groups of men who were in the judges' confidence. In return for a steady stream of bribes, guardians acquired absolute control over the resources of Indian minors, which were then funneled toward various corporate officials. The system allowed whites to milk oil-rich wards to line their own pockets and those of their cronies and the oil companies that employed them.[53]

Those who fared the worst in the face of organized and institutionalized discrimination were those who were themselves unable to organize effectively.[54] Tenants, blacks, labor unionists, and Socialists, among others, failed to wield organizational clout. As a result, they suffered inequities and were subject to violent attack with little recourse to justice. For Indians, whose resources had been a magnet to which the corrupt were drawn, disorganization was a dangerous tendency. One did not have to look long to find abundant evidence of how easily Indians were swindled of their resources and how they lay exposed to the demands of others.

Disillusioned by the mounting criminality in Oklahoma, progressives began to turn toward their fellow tribesmen. Indeed, this was an idea that was gaining credence both within and outside the tribes. In 1925, Malcolm MacDowell, secretary of the board of Indian commissioners harangued an assembly of progressive tribal members: "What are you doing for your children?" "What are you leaders among the Indians doing to help these fellows who live off in the hills and do not know how to take care of their affairs?" "I have heard eloquent Indian orators tell how their hearts bled for their people, yet would not go across the street to learn whether a little Indian child is in school."[55] Progressives, though, had already begun the process of reconciling with more conservative-minded tribal members. Beyond the work of those tribal members employed by the Indian bureau, progressives began to engage in activities that bespoke a need to gather as, and enact change for, Indians.

The form of these efforts varied. Like-minded progressives in urban areas founded social clubs and civic organizations along the lines of other educated middle class people. In the mid-1920s many Indians in Tulsa participated in the Apela Club. This group, with a membership upward of one hundred tribal members, including some intermarried white men, met for lunch each week to discuss issues of significance to Indians in the city. Made up of various businessmen and professionals, the Apela Club was an urbane organizatio[56] dedicated to civic good works, like planting trees at the state fairgrounds.[57]

But it was also more than that. Apela was a Choctaw word meaning "help," and the group's members promoted Indian causes. They raised money for scholarships for Indian children and raised consciousness about traditional Indian practices. Its membership, although open to intermarried tribal members was reported by one newspaper as being restricted to Indians by blood. "No eagle feathers grace the brows of its members, and no tom-tom beats foretell its weekly meetings," a reporter wrote, but the club was an expression of Indianness, however unlikely.[58]

Advocates of Oklahoma's development since the late nineteenth century, progressive tribal members realized that the temporary "pain" caused to conservative tribal members as a result of allotment had become permanent, had become institutionalized. They also understood that they themselves were subject to the vortex of misery and discrimination that victimized so many tribal members. As the historian Angie Debo has noted, "*all* classes of Indian citizens" suffered financial losses by the 1920s"[emphasis mine].[59] For progressives, the creeping dangers of ruthless oil speculation, withdrawal of federal protections, declining economic conditions, and rising intolerance shattered the hierarchies of difference within the tribes. The attitudes of whites toward Indians had grown, in many respects, less discriminating over the period of Oklahoma's territorialization and early statehood. White society increasingly failed to register distinctions among tribal members, preferring to lump them together as inferiors. Indians, it was clear, would simply be left behind by whites unless Indians took action. The Apela Club, by focusing on Indians and Indian concerns, addressed matters that other groups and local government ignored.

* * *

While only a small example, the Apela Club signaled the beginning of a rejuvenated Indian activism. Indeed, the hesitant steps by progressives to join with conservative tribal members advanced steadily through the 1920s. Fittingly, Victor Locke, Jr., an accomplished and ambitious progressive, heralded the arrival of a larger, more politicized movement in the aftermath of his dismissal from the superintendency of the Five Tribes. In 1924, Locke convened a meeting with Indian representatives from around the state. As Angie Debo has sketched the scene, Locke sat down with "blanket" Indians from the western reaches of the state, men bedecked with feathered headdresses, and passed a pipe of friendship.[60] By sitting with them, appealing to them, Locke hinted at the shift taking place in the way diverse Indians addressed the knotty historical problems of their own factionalism. In the coming years, as older strategies of resistance and assimilation appeared worn out, Indians gathered to formulate new responses that focused not on their differences but on what they shared.

SEVEN

"Indians, Organize!" A Peoplehood in Blood

In 1930, Rachel Caroline Eaton, a Cherokee, wrote an historical account of an early-nineteenth century clash between the Cherokees and the Osages, an event later known as the Battle of Claremore Mound. Pushed westward by whites, the Cherokees found themselves on the periphery of "civilization" in western Arkansas. There they endured constant harassment from the Osages who took a dim view of the Cherokees' trespass on their lands. Eaton detailed the events leading up to the conflict and the steadily escalating pressures on the Cherokees to defend themselves against Osage hostility. Horse-raiding parties routinely plundered the Cherokees and ambushes were frequent. Finally, outraged by the deaths of several tribal members in an engagement with Osage warriors, the Cherokees prepared to fight back. After mobilizing their collective resources, they tracked down their tormentors and routed their war-hardened foes.[1]

Although Eaton made clear that the Cherokees benefited from coordinated battlefield tactics and advantages such as skilled marksmanship in defeating the Osages, she argued that their success derived primarily from other sources. Her essay focused on the Cherokees' ability to work together in quest of a common goal. Eaton detailed the tribe's cooperative preparations for war, single-minded planning for battle, and sublimation of personal desires to those of the group. Bending their wills toward a shared end, the Cherokees produced an astonishingly total victory. In fact, it was less a battle than a massacre; the Cherokees had nothing to fear from the Osages afterward. Cherokee unity, Eaton argued, ensured their survival as they contested for land in the West.

By the 1920s and 1930s, such lessons resonated strongly among members of the Five Tribes. As land continued to slip from their grasp and the state pushed unrelentingly to terminate remaining tribal autonomy, a number of authors, harkening to examples of older achievements, emphasized the necessity of tribal unity in the face of outside threat.[2] These ideas extended beyond the literary elite. Tribal writers merely articulated the thoughts of the many men and women who were struggling to fashion a unified politics among the Five Tribes. Recognizing that factionalism had served them badly, they courted solidarity. One tribal member, taking note of the troubles and disillusionment suffered by his fellows, declared: "This is a day of organization. The Indian must organize. . . . Living as we are today in a state noted for its rotten politics, its corrupt public officials, its lawlessness and its racial and religious intolerances Indians, organize!"[3]

Such sentiments were shared widely within the tribes. Both progressive and conservative tribal members acknowledged that the differences among them need not divide them. Recognizing the limitations of assimilation, progressives advocated a re-strengthening of tribal structures and offices. Concerned that equality within American society was blocked by corrupt whites and disinterested institutions, they participated in a surge of organizational activity. Calling for a recognition of Indian rights and a commitment to the defense of Indian peoples, progressives held formal conventions, published periodicals, and pressed official resolutions upon the government. At the same time, conservatives confronted the shortcomings of their previous strategies and began to abandon resistance and separatism. They instead supported a variety of grass-roots movements that promoted cooperation with fellow tribal members. On the whole, tribal members breathed life back into the idea of the tribe as a flexible structure, one that could accommodate a range of beliefs and behaviors.

In their bid for unity, tribal members dealt with their former internal enmity. Separatists retreated from their rigid criticism of opponents for their lack of legitimacy as Indians. Their claims of "real" Indianness subsided and they began to accommodate their fellow tribesmen and even intermarried whites. Progressives, meanwhile, grappled with their heritage and their obligations to other tribal members. As the abuses of whites mounted and as federal support wavered, a number of progressives identified more readily with the Indian elements of their ancestry. Although they continued to pursue professional ambitions and upward mobility, they stepped up efforts to improve the lives of their fellow tribesmen.[4]

As tribal members focused less upon their differences and more upon their similarities, they began to reconsider their previous uses of blood. Blood continued to form a vital part of identity. But now, instead of dividing tribesmen from one another, ancestry became a common ground. A new and inclusive sense of peoplehood arose that revolved around blood. Importantly, blood quantum diminished as a measure of Indianness.[5] Shared bloodlines, not restrictive definitions of quantum, held out the possibility that the tribes could retain their distinctiveness while effectively advancing their interests.[6]

* * *

Shortly before his death in 1917, Redbird Smith, the charismatic leader of a band of Keetoowahs called Night Hawks, expressed grave doubts about the path he and his followers had chosen.[7] Having branched off from the legitimate Cherokee government in the waning years of Indian Territory, Smith reflected critically upon the separatist strategy he had advocated so passionately. According to his confidant and longtime Keetoowah official Levi Gritts, "[b]y the time Redbird Smith died, he told his people he believed he had taken a wrong course and advised them that it was never too late to correct a mistake."[8] Gritts, writing in 1930, explained that Smith regretted that disagreements among the Cherokees accelerated the tribe's fragmentation and decline.

In the years immediately following Smith's death, various groups of Cherokees worked to effect unity. In November of 1920, the Illinois Fire of the Keetoowah Society, one of seven society "fires," drafted a statement concerning tribal affairs past, present, and future. The Keetoowahs noted that they had been "overwhelmed" by the loss of land and sovereignty in the last years of the nineteenth century and first years of the twentieth. Stripped of "what they considered their vested rights and prerogatives of self-determination in the National Government affairs," the document observed, "they [the Keetoowahs] crystallized into a recalcitrant and combative mood."[9] Their primary enemy became the federal government, and the society sought to distance itself from the Indian bureau. But the Keetoowahs did not only "look . . . askance and with skepticism" at the government's pronouncements and actions. They also broke with other tribal members. The "Cherokee people," subjected to exploitation by white society, "became factionalized and bitterly antagonistic towards one another," the resolution maintained, "rendering impossible a united Cherokee effort for mutual benefit." Unfortunately, the Keetoowahs noted, this state of affairs only served to abet "the unscrupulous and exploiting hordes who infested our country and plied their graft."[10]

Seeking to reverse its former defiant stance and move toward more constructive strategies, the Illinois Fire offered a revisionist re-telling of the recent past. The post-statehood years, the Keetoowah statement declared, had proven "salutary" for the Cherokees. Influenced positively by the arrival of industrious whites, "home builders and hence nation builders," Cherokees began to take their place among the leaders of the new state. Through "decades of close contact and living in the atmosphere of intensive and constructive effort . . . all attending factors [of Cherokees were brought] into closer harmony."[11] Perhaps no more important event drew the diverse segments of the Cherokee population together than World War I. During "our recent World Struggle," the Illinois Fire noted, the United States "called upon its whole people for a united effort. To this all, not only the full-blood and mixed-blood Cherokees, but all the American Indians gave a spontaneous and unequivocal response, placing their manhood and material assets at the disposal of 'Uncle Sam', thanking him for the opportunity to show the world what is inherently in them as a race."[12] As a result of their participation in the war, the Keetoowahs wrote, tribal members developed a "new psychological angle." Convinced for the first time that they had become "part and parcel of the body politic . . . and not an isolated and segregated 'Indian problem,'" they were "ready to re-assert their self-respect and confidence in themselves."[13]

Yet these changes did not suggest that the Cherokees should or could melt into the mainstream. Rather, tribal distinctiveness was to be preserved. The best way to accomplish that goal was to reverse the former exclusionary outlook of the Keetoowahs and acknowledge that many types of Cherokees shared tribal membership. "The unrestricted, mixed-bloods, and the competent . . . are equally interested and have equal rights" in tribal affairs, the Illinois Fire statement noted. The Keetoowahs were "sensible of the prevailing spirit of cooperation among the different groups and organizations of our people, they are alert to the opportunity at hand for effective cooperative work."[14] There were significant material considerations motivating the Keetoowahs. The Illinois Fire noted that tribal members needed to work together in order to prosecute outstanding claims cases that were working their way through the courts. These claims, for lands lost in the Southeast as well as during the allotment period and land rushes of the late nineteenth century, held out the promise of significant per capita payments to which all tribal members had an equal stake.[15]

The Keetoowah account of developments dating from the Dawes reforms to 1920 underlined the necessity of leaving resistance and separateness behind and forging new alliances. The group proclaimed, in explicit terms, its willingness to correct its course. The Illinois Fire was not alone in this effort. Apparently following their former leader, Redbird Smith, members of the Night Hawk Keetoowahs made similar overtures of reconciliation, first to other Keetoowahs and then to the general Cherokee population. In 1921, the Night Hawks convened with the larger body of Keetoowahs at a meeting in Tahlequah, the capitol of the old Cherokee nation, in order to salve the wounds that had rent the organization in past years. Present was W. T. Brady, described in press reports as a "prominent businessmen in Tulsa," who commented on the union of tribal members. After decades of strife, it was "possible for Cherokee citizens of all shades of political and religious belief to come together in an indissoluble union," Brady maintained.[16] *The Tahlequah Arrow-Democrat* commented on the "happy union of all classes of Cherokee citizens"; various tribal members and organizations "now assemble together in friendship as citizens of a United Cherokee Nation, all imbued with the desire to work together for the interests of all and in harmony."[17]

The Night Hawks also exhibited other signs of their waning enthusiasm for separatism and resistance. In 1918, they recruited an Oneida attorney named C. P. Cornelius to advise the group. Picking up the mantle of leadership from the deceased Redbird Smith, Cornelius embraced several ambitious plans. He petitioned the Indian bureau to lift restrictions on several Keetoowah-held allotments in order that the individuals might purchase tracts adjacent to one another. The idea was to foster a community in which older practices of sharing and cooperation might continue. Then, with the proceeds of the land sale, Cornelius started up a bank to finance such projects as the raising of a group-owned cattle herd.[18] Reporters were also periodically invited to visit the Keetoowahs in an effort to convey the non-threatening and industrious nature of the community. Famously guarded in their membership and activities, the Night Hawks' openness suggested a departure from older tactics.

The changed views of the various Keetoowah groups rippled outward and began to affect the affairs of the larger body of Cherokees. While the Tahlequah meeting of 1921 witnessed the coming together of Keetoowah factions, it also welcomed other Cherokees to participate in an election that had as its object the revitalization of Cherokee government and the reunification of a people badly divided. Given that Cherokee leaders were no longer elected by the tribes, but rather were appointed by the President, this was in itself a bold move. Some four hundred Indians turned out and press reports noted that the Cherokees in attendance represented various groups and classes of tribal members. Among their motivations, according to a newspaper report, was the belief that Cherokee claims cases would benefit from "a supreme head or authority" of the tribe.[19] This diverse assembly joined in support of Levi Gritts as the nominee for the office of principal chief. Described as a "man . . . whose integrity of thought and regard for the various groups of our people are at all times compatible with their varied interests, and a man endowed with a native-born pride and ambition for his people," Gritts became a vehicle for the forces of unification.[20]

In October 1921, despite the refusal of the Department of Interior to sanction the tribe's election, Gritts elaborated on the import of the Cherokees' actions and the changes underway in their approach to tribal unity. "'My people,'" he was quoted as saying in a newspaper article, had "'reached a point where harmony had disappeared and where prejudice existed among themselves so strongly that for several years their condition became horrible not only with reference to their property holdings, but they were losing their self-respect and confidence in one another; but the time has come now that they have realized their mistakes.'" After recounting the motivation behind his election, Gritts stated:

> We have plans and undertakings which will without doubt be successful and we may at some future time be instrumental in helping our brother Indians. . . . But we realize that the first and most important thing for us to do is to organize, so we can understand one another and all work for the same cause. We realize when we organize that we will need to educate our

children, remind them of their responsibility to God, teach them to be true lovers of their people as well as their white brothers, and teach them to be industrious in all lines of business.[21]

The former views of the Keetoowahs, their often combative approach to white society, appeared to be in eclipse.

In a letter to Victor Locke, Jr., who was at the time serving as Superintendent of the Five Tribes Agency, Gritts further explored these ideas. The founding of Oklahoma and its subsequent rush of development, he wrote, had forced the Cherokees to devise and adopt a new political course.

> We know that our nation as a political organization has ceased to exist and that it can never be restored; a changed condition in this country has made it impossible for the Indian longer to maintain his old and separate form of government. . . . The Indian does not even dream of restoration.

Yet despite these harsh conclusions regarding past decline and present frailty, Gritts wrote with forward-looking purpose. The Cherokees do "not wish to contemplate the time when it must be said, 'There is no Indian.' . . . We desire the freedom of action in the matter that will most nearly accomplish the purposes we have in view." At the heart of these purposes was the "perpetuation of the Cherokee Indian in some form of a political or legal organization by which they may be tied together." Although they had lost the Cherokee nation as it previously existed, they retained a "nationality" that was inclusive in its definition. When Gritts mentioned his election as chief, he allowed that he represented not only specific factions within the tribe but all of "the forty thousand some odd Cherokees."[22]

Given impetus by a separatist organization, the movement among the Cherokees suggested a broad base of support for a new political direction. Openly acknowledging the "different factors of Cherokees," and "fully cognizant and sensible of the social and political position of the unrestricted Cherokees" who have "no further need of the good offices of a Chief, except in-so-far as it facilitates the cooperation and coordination of all Cherokees in the protection and prosecution of interests and claims in which all have had mutual and equal right," the Keetoowahs received a significant response from their fellow tribesmen.[23] The ties that linked tribal members together appeared resilient despite the differences within the tribe. And the potential benefits to tribal members, from claims cases and other activities, appeared greater through unified action.

Among the Chickasaws, similar developments gathered momentum throughout the 1920s. In an effort to work in concert with and influence the Chickasaw Governor, Douglas Johnston, and to ensure that their priorities gained an equal hearing in Chickasaw affairs, a number of conservative tribal members residing in the vicinity of Seeley, OK., created a grass-roots movement that expanded throughout the decade. The Seeley Chapel Association began in the early 1920s before giving way to the Seeley Indian Association shortly thereafter. As the ranks of its members grew, the group later spawned the Chickasaw Protective Association as well as an inter-tribal organization that sought to work with the Choctaws on matters relating to the resources shared by both tribes.[24]

Confronted with unfavorable legislation by state and federal authorities, these conservative-led groups attempted to broaden the strength of the tribe to preserve what resources and protections they still retained. For example, when talk of re-opening the tribal rolls circulated among officials, the Seeley Indian Association spurred Governor Johnston to protest.[25] While mistakes had been made by the Dawes Commission in enrolling tribal members, and numerous individuals had been wrongly left off the census lists, the Chickasaws considered the idea of re-opening the rolls anathema. They feared the further dilution of land and mineral rights and viewed the effort as a guise to allow whites even greater access to Chickasaw holdings. The threat was a clear one, and tribal members understood what they stood to lose if the rolls were re-opened. In response to pleas from the Seeley Indian Association, Johnston, who also did not favor the prospect of having the tribal census revamped, supported legislation "keeping the rolls closed against thousands of claimants" for Chickasaw citizenship.[26] At a Chickasaw convention in 1924, he maintained that large per capita payments which tribal members had received in the past resulted from "purging the rolls of approximately four thousand Court Citizens who had been admitted to Citizenship by decree of the United States Court."[27] Court citizens were those who had been included on the Dawes Rolls at the request of the courts that reviewed the citizenship lists, and Johnston implied that further tinkering with the rolls might jeopardize the tribe's assets and thus the welfare of individual tribal members.

Beyond the Dawes Rolls issue, the Chickasaws demonstrated the emerging political partnership of conservatives and progressives in their efforts to secure the extension of federal restrictions on their allotments. By 1927, the Seeley Indian Association was working with Johnston to lobby Congress on behalf of Indian interests. Forbis Cravatt, a conservative leader, corresponded frequently with Johnston on the project and was assured by the governor that the Chickasaw national attorney had "talked the restriction matter over with the Commissioner, who is in favor of it, and Mr. Carter [an Oklahoma congressman and a Chickasaw] wanted to take up the matter before he left Congress."[28] In a letter to Congressman Scott Leavitt, who chaired the House Committee on Indian Affairs, Johnston discussed the concerns of conservative tribal members and their stance on legislation favorable to their interests. A number of Chickasaw tribal members, he wrote, "have passed resolutions in conventions memorializing the Department of the Interior and instructing their tribal representatives to secure legislation extending the restrictions, as they realize and look forward with dread to the condition that confronts them at the expiration of the present restricted period should adequate legislation not be enacted for their protection. They know what would happen—that they would soon be homeless and in destitute circumstances."[29]

By the following year, 1928, the Chickasaws undertook inter-tribal organization with their neighbors, the Choctaws. The object of talks between the tribes was to stave off taxation on their lands, and in March 1928 there was a well-attended joint Chickasaw-Choctaw convention at Ardmore, where "probably 250 or 300 . . . Indians [were] in attendance and from almost every portion of the Choctaw and Chickasaw Nations."[30] A correspondent of Gov. Johnston noted that "[t]he real reason why the convention was numerously attended was because of the approach of taxation—they realize that unless money is paid to them by the government they will probably lose their lands—that is unless the taxation time is extended and they all know that if extended it will only apply to a small part of their present homestead."[31]

Confronted with the insistent attacks of state officials and the neglect of the federal government, the Chickasaws worked together to promote their varied interests. Conservatives no longer considered separatism a viable option, but preferred to keep a close eye on the men who had nominal authority over tribal affairs. And while unrestricted mixed bloods were not personally interested in the extension of restrictions governing allotments, they nonetheless agreed that a common stand was necessary in order to promote a full range of tribal political objectives. From this pragmatic approach to tribal affairs there derived a more indulgent view of tribal membership; a recognition that differences among tribal members need not result in intractable factionalism. Indeed, there was explicit recognition that all Chickasaws enjoyed a shared bond. When the Chickasaw Protective Association was created in 1929, an "executive committee" established guidelines for representation.[32] "All duly enrolled Chickasaws and their descendants twenty-one years or older are eligible to membership," read a resolution on the issue.[33] In a further clarification, it was stated that the convention had been called in the name of the citizens of the Chickasaw Nation (full blood, mixed bloods and intermarried)."[34]

Similar undertakings occurred in the other Five Tribes. Among the Seminoles, for instance, Charles Wisdom, an anthropologist sent by the government to study the tribe in the 1930s, noted that while tribal factionalism remained in evidence within the tribe, roughly ninety percent of tribal members endorsed one political group. Headed up by young, progressive Seminoles, the party's supporters spanned class, culture, and ancestry lines. Most tribal members were attracted by the political demands made by these young leaders, which included renewed tribal sovereignty and increased access to governmental loan programs available through the Indian New Deal.[35] "This is the only faction," Wisdom wrote, "which thinks completely in terms of the entire tribe, rather than merely in terms of its small group interests, and it is the only one genuinely interested in tribal welfare."[36] In addition, a central idea that helped to keep the party together despite the differences in its supporters' backgrounds was a fierce anti-black attitude. Its leaders called often and loudly for the exclusion of Seminole freedmen from participation in tribal government and from sharing Seminole assets.[37] They were, according to Wisdom, "most violently oppose[d] [to] the freedmen in any tribal set-up."[38]

The inclination of conservatives to solidify a broad-based politics within the Five Tribes was mirrored by progressive endeavors. Occurring during the same period, progressive groups such as the Apela Club sprouted up and began to carry out campaigns to improve conditions for Indians and to erect defenses against their decline. Perhaps the most important of these organizations was the Society of Oklahoma Indians, which was founded in Tulsa in 1924 in response to a call by progressives to promote Indian interests across the state. The Society set out to define the major problems confronting the state's Indian population. Its objective, as stated on the cover of the booklet printed for the first annual convention, was "to protect the civil, social, educational and financial rights of Indians of the State of Oklahoma."[39] These rights, according to several of the speakers who addressed that first convention, were the full due of "the Indian people."

Running beneath the rhetoric that justified the launch of the Society, however, was the understanding that Indian survival depended upon Indian solidarity. In February 1924, at the Society's first annual convention, some four hundred Indians gathered in Tulsa, determined that their best hope was to band together. As a single group, they might better stem the tide of dispossession and discrimination. S. R. Lewis, a Cherokee and chairman of the convention, stated: "This is a day of organization. The Indian must organize. . . . Living as we are today in a state noted for its rotten politics, its corrupt public officials, its lawlessness and its racial and religious intolerances Indians, organize!"[40]

Lilah Lindsey, who claimed both Creek and Cherokee ancestry, noted the importance of presenting Indian interests through collective action. "We should know each other better, and stand up for our rights and see that our rights are preserved. Let it be our slogan that we are Indians organized for the protection of the Indian."[41] Other speakers urged Indians to put aside past differences. They maintained that factionalism, whatever its nature, only weakened the tribes. Sen. E. M. Landrum of Vinita remarked that, "[w]e should get together on a common ground to defend our rights together."[42] W. T. Brady added that "[i]t is time in the history of our great state that the Indian people, regardless of politics, regardless of tribes, stand together for those things that are due them."[43]

The invitation to heal old wounds was broadly articulated and the inclusive tone of the society's annual meetings was established in symbolic and actual terms. The conventions were vivid affairs attended by urban professionals and small farmers alike. Until the Society petered out in the late 1920s, events such as golf tournaments and spectacles such as full blood "encampments" were advertised at annual gatherings. The encampments, replete with authentic, traditional lodgings and food preparation demonstrations, contributed to the growing sense of solidarity that the Society espoused.[44] Additionally, conservative leaders were issued formal invitations to lecture and to serve on the various committees that drafted resolutions. The resultant mingling of diverse individuals and viewpoints underscored the notion that all tribal members shared in the endeavors at hand.[45]

Various convention speakers sought to establish a basis upon which diverse Indians might define and promote a shared sense of Indianness. In so doing, many offered commentary on the utility and flexibility of race in defining Indianness. In groping toward an understanding of Indian distinctiveness within the larger society, participants at the 1924 Tulsa meeting promoted their ancestry. C. P. Cornelius, the Oneida who counseled the Night Hawk Keetoowahs, spoke openly about racial pride. "One reason for being proud of the Indian race is the fact that this is one race who has maintained their characteristics through nearly a thousand years association with a foreign race. . . . We have never showed anything but the most magnanimous feeling toward other races. We have learned that all men are created equally and it is up to each individual that he make the most of his race."[46]

Others, while no less adamant of Indian distinctiveness, were not as generous as Cornelius. In fact, some speakers at the Society's meetings championed Indians in opposition to other minority groups. When Fannie Starr of Muskogee took the podium in 1924 at Tulsa, she admitted to being "almost too prejudiced to make a speech for I am all for the Indian." Although not a tribal member herself, "[m]y father was interpreter for the council. I love to see the Indian faces and I love the cause of the Indian." She then added, "[w]hat we want now is the nigger money for the nigger land."[47] The Society, and other groups, worked throughout the 1920s for the return of freedmen lands or compensation for those lands; alone among the defeated populations in the Civil War they argued the Five Tribes were punished by having to give their former slaves land. Indeed, speaking after Fannie Starr was Col. P. J. Hurley, a former Choctaw Nation attorney, who recounted that in his youth public education had been denied to the Indians "unless he degraded himself by going to school with the negroes."[48] In the future, he argued, Indians should not be so debased.

Discussions of race also touched on the sensitive subject of mixed bloods and intermarried whites. After W. T. Brady called for Indians to unite, the Congressman added that "I am proud . . . to be a member of the inter-married citizens of the Cherokee Nation. I have five children and four grandchildren of Cherokee blood." While acknowledging that his own membership in the tribe was subject to legal standards established by the Dawes Rolls, Brady suggested that his marriage and the birth of his mixed blood children strengthened his ties to the tribe and to his fellow tribal members.[49] Given the apprehension that whites were insidiously marrying into the tribes solely to obtain access to valuable lands and mineral rights, Brady's plea reminded tribal members of the usefulness and security of a more inclusive approach to Indianness.

The tendency to appeal to shared Indian interests was reinforced in the official acts of the Society of Oklahoma Indians. Through a flurry of resolutions and petitions, the organization attempted to voice its concerns to audiences in both Oklahoma and Washington, DC. In particular, it weighed in on the battle between state and federal officials over the control of Indian affairs. Members of the society worked to bolster the federal government's continued presence in Oklahoma, and it frequently called for "a spirit of cooperation" between Indians and the Indian bureau. Numerous resolutions requested the preservation of Indian lands and the need for federal protections. One resolution argued that the government was vital in helping certain Indians in the "conservation of the estates of our people as the primary principle of self-maintenance and future independence." The resolution added that "[w]e are looking down the years, not only of the lives of the present generation, but as well of the lives of the generations to come." Although these were the supposed fruits of assimilation, the society understood that only through the continued attention of the federal government might "self-maintenance and future independence" be achieved.[50]

Progressives, having grown disillusioned regarding the prospect of assimilation, and apprehensive about the government's faltering support for Indian programs, labored alongside conservative tribesmen to ensure that a federal buffer would remain in place to stave off the depredations of white society in Oklahoma. Concerns about education were aired at the 1926 convention, and a state official was criticized for stating that Indians who lived on untaxed lands could not be educated in Oklahoma's public schools. "[T]he Indian children of the State of Oklahoma are entitled to all the rights and privileges of other children," the Society maintained. "[I]n our opinion, we consider such officials [as the one who threatened to deny tribal members access to public schools] to be prejudiced against the Indian race."[51] The Indian bureau was called to task in negotiating better mineral leases and ensuring honest probate practices in Oklahoma. Regarding corruption in Oklahoma, numerous speakers at Society meetings provided strong testimony, and the sensational report issued by the Indian Rights Association, "Oklahoma's Poor Rich Indians," was handed out to each participant at the 1924 convention to underscore the gravity of the conditions confronting tribal members. Other resolutions had as their object the continuation of various claims cases that were designed to elicit payments both for lands forfeited during the removal era as well as those lost in the wake of the Dawes reforms. Additionally, a number of resolutions were specifically written on behalf of the Five Tribes. During its annual convention in 1925, the Society requested that "no full-blood Indian of the Choctaw, Chickasaw, Cherokee, Creek or Seminole Tribes shall have power to alienate, sell, dispose of or encumber in any manner any of the lands allotted to him for a period of twenty-five years."[52]

Despite the congenial tone advocated by the Society of Oklahoma Indians in its dealings with the federal government, progressive tribal members frequently went beyond the organization's bounds and engaged in direct attacks upon the Indian bureau. There were even instances when conservatives joined with progressives in denouncing government mismanagement and even corruption. Both groups complained about Indian bureau employment practices. Many tribal members considered the federal government an important resource for jobs. It was one of the few institutions that supposedly embraced an open policy in hiring Indians. For progressive tribal members especially, the Indian bureau provided a place where high-level positions were available. They considered employment with the government to be a particularly important proving ground—for both Indians and the government. Yet in a number of instances, government hiring choices incited tribal members' anger—and another reason for progressives and conservatives to band together.

In 1925, a Cherokee named O. K. Chandler, who had served as superintendent of the Quapaw Agency in northeastern Oklahoma, criticized then Commissioner of Indian Affairs Charles Burke and Five Tribes Agency Supt. Shade E. Wallen. The charges leveled by Chandler were several, but foremost among them was of a general attitude of hostility toward and discrimination against Indians. He was especially vocal about the failure of the Indian bureau to hire more Indians, particularly to serve in high office.[53]

In late summer of 1925, Chandler visited numerous groups among the Cherokees and Creeks. He distributed cards on which were printed anti-Indian phrases attributed to bureau officials. "Are you an Indian?" the card asked, followed by the statement, attributed to an Indian bureau inspector: "A damned Indian is not good enough to be superintendent of an Agency." The card maintained that this was hardly the utterance of a lone official, but that it summed up the philosophy of the entire Burke administration. The card noted how Indian school principals as well as agency officials were routinely replaced by whites. Men "not of Indian blood," the card emphasized, were those most likely to obtain high positions in the Indian bureau administration.[54] During his visits with tribal groups, which one correspondent derogatorily referred to as the "Chandler meetings," Chandler recounted his charges and provided resolutions to be voted upon. Not only were adult Indians discriminated against, he maintained, but Indian children were receiving poor education at the hands of whites. And even if they did receive a sound education, "it didn't make any difference," because they "could not get a job in these Government schools" as teachers or administrators. Moreover, the Burke and Wallen administration was mishandling Indian resources and legal claims.[55]

In August 1925, a gathering of Keetoowah Society members endorsed Chandler's charges and submitted a resolution to have Burke dismissed from office. A later Keetoowah meeting, in October 1925, endorsed the Burke and Wallen regime, but it did so conditionally. Attempting to use what leverage they could, the Keetoowahs tacked on to their resolutions praising the bureau commissioner and Five Tribes superintendent several demands. There existed, the Keetoowahs stated, an "urgent need of Farm Demonstrators, as well as Demonstrators in Home Economics and Visiting Nurses" and they requested that funds be forthcoming to supply personnel.[56] Although Chandler and the Keetoowahs did not seek the same ends, they nonetheless found one another useful in pressuring the Indian bureau.

For the government, which went to significant lengths to discredit its former Quapaw Agency superintendent as a disgruntled crank, Chandler was a serious nuisance. One detailed report submitted to a bureau inspector by a probate clerk showed that following Chandler's visit to a group of Keetoowahs the bureau sought out and interviewed each signer of a resolution protesting the Burke administration. The report was replete with the time and location of each interview. A number of those interviewed at their homes by a federal official disavowed their actions. Several tried to pass the buck to others and suggested that they did not harbor anti-government sentiment.[57]

In response to the Chandler meetings, Wallen began to host feasts and other events where he could address complaints made against the bureau and win support. These tactics aggravated at least one Cherokee group at Moody, OK., which passed resolutions condemning Wallen and the bureau. "Wallen and his associates," the *Muskogee Daily Phoenix* reported, are accused of holding barbecues and meetings at which Indians are asked to sign prepared resolutions endorsing his own administration and that of Commissioner Charles H. Burke."[58] One of these resolutions, citing Wallen's "experience and intimate knowledge of the business interests of the Indians," was signed by "The Full Blood Members of the Creek Tribes in Mass Meeting Assembled" at Okmulgee in May 1925.[59] The *Daily Phoenix* went on to note that the superintendent was also criticized for "seeking through newspaper publicity to draw a distinction between mixed-blood and full-blood Indians 'with the apparent purpose of bringing about discord.'"[60]

The charges and counter charges mounted through 1925, and when the Society of Oklahoma Indians met at its annual meeting in 1926 in Muskogee, a large anti-Burke contingent was in attendance and made itself heard. The recent free-wheeling criticism of the Indian bureau spawned an aggressive atmosphere among those in attendance and led to resolutions on various matters, many of them condemnatory in tone and content.[61]

The undertakings of Chandler and the society of Oklahoma Indians, as well as various grass-roots movements among the Five Tribes, suggest a reawakening of political energies. Decrying Indian victimization, and calling for solidarity in the face of outside threat, tribal members extolled the virtues of Indians and the potential for collective action. In key instances, diverse tribal groups dismissed factional antipathies in favor of cooperation.

The emphasis of tribal organizations and individuals on achieving a unified politics was echoed in much of the writing produced by the Five Tribes during the 1920s and 1930s. This was especially the case with those who set out to re-write tribal history. According to the scholars Daniel Littlefield and

James Parins, there was a "tremendous literary energy and extensive production of a rich literature, which blossomed in the late nineteenth century and bore fruit in the twentieth."[62] In what Littlefield refers to as the "reform era," or the first decades of the twentieth century, "[h]istorical narratives and biography" were particularly important forms of literary expression among members of the Five Tribes. "Knowing that their nations were passing through a dramatic time in their history," he writes, Indians "realized the importance of writing tribal history long before it became fashionable among non-native historians to do so."[63]

The lessons of the past were central to the conception of unity espoused by tribal members. Just as Rachel Caroline Eaton depicted "The Legend of the Battle of Claremore Mound" with a keen eye for telling details of collective action by the Cherokees, so other writers sought out instances in which tribal members succeeded through cooperation. In a 1932 essay on Stand Watie, a Cherokee who was the last Confederate general to surrender to the North, Mabel Washbourne Anderson used biographical investigation to underline the necessity of presenting a unified front in the face of danger.[64]

Anderson's praise of Watie focused on his ability to look beyond partisan politics and work for the best interests of the entire tribe. Even after his brother was killed by tribal foes angered at his policies, Watie continued to promote the welfare of all Cherokees. "Unshaken by feuds and factions," Anderson wrote, "which constantly threatened his life, from that time on his power, purpose, and courage proved of lasting influence."[65] Accordingly, wrote Anderson, he worked for a treaty of unity that was signed in 1846 among the various tribal factions. The result was a period of "peace and prosperity for the Cherokees." Watie continued to participate in public life, despite his wish to pursue the "home life he loved."[66] In so doing, he demonstrated selfless leadership.

Even after events spun out of control and Watie advocated alliance with the South, other Cherokees respected his motivations and his abilities. While he considered himself a Southerner, he strove to keep the tribe intact. In battle Watie proved himself time and again a brilliant tactician and courageous officer. Anderson wrote that mixed bloods, white intermarried Cherokee citizens, and full bloods alike were unstinting in their praise of Watie. So great was Watie that he held out the potential of unifying the diverse segments of the tribal population.[67]

Tribal writers also wrote for a non-Indian audience, and in doing so they sought somewhat different objectives. As the federal government reduced funding for Indians and withdrew support for education and other important programs, writers attempted to combat the downward spiral of Indian life by appealing to a sympathetic white audience outside of government. In particular, they sought to engage a fervent reform movement that had sprung up in response to the corruption and mismanagement that marked the Indian bureau and tragic decline of Indians from around the nation. This movement was comprised of intellectuals, philanthropists, artists, and others who possessed the wherewithal and the will to mount effective critiques of federal policy and to wage battles to reverse plans harmful to Indian groups. Such was the reformers' energy that the 1920s, according to the historian Francis Paul Prucha, "was a time . . . of widespread and bitter criticism of the Indian Office, its officials, and the program they promoted. The cry for reform was in the air and it grew in volume as the decade passed."[68] Organizations such as the Indian Rights Association, the Board of Indian Commissioners, the Indian Defense Association, and the Indian Welfare Committee of the General Federation of Women's Clubs kept the heat on government officials in an effort to improve the conditions of Indian life. Their members lobbied politicians and published articles in various popular and scholarly journals in order to correct what they viewed as mistakes in Indian policy.[69]

These reformers labored to improve programs for Indian education, health, and economic development. They launched numerous investigations to expose governmental corruption and the poor handling of official responsibility toward Indians.[70] As a result, they played an important role in several crucial events of the mid-1920s. Scholars credit Indian reformers, including John Collier, who later became Commissioner of Indian Affairs under Franklin D. Roosevelt, with helping defeat the Bursum Bill, which sought to undermine Indian land titles in New Mexico. In a campaign that attracted national attention, Collier led an unrelenting and at times personal attack on the Indian bureau and the Department of the Interior. Joined by other influential activists, reformers successfully beat back Secretary of the Interior Albert Fall and then helped cement their victory with passage of the Pueblo Lands Act which introduced a plan to preserve Indian lands.[71] Elsewhere, reformers pushed for legislation to protect Osages from unscrupulous guardians. The Osage Guardianship Act of 1925 granted greater power to the federal government to safeguard Osage land and mineral rights against the designs of Oklahomans.[72]

These reformers were susceptible to "romantic" images of Indians, and their crusade on behalf of the continent's original inhabitants was marked by passionate devotion to cultures that appeared to be on the brink of extinction. Indian subjects cropped up with rising frequency in the fashionable salons of figures such as Mabel Dodge, the philanthropist and activist. When Dodge and others moved from New York City to Taos, New Mexico, a thriving artists' colony, a fuller exploration of Indian life followed. The nearby Taos Pueblo and the natural beauty of northern New Mexico inspired a number of the new arrivals to re-examine their ideas of Indians. Writers such as Mary Austin looked to Indian music as the source of a distinctly American poetry. Indian pottery, blankets, baskets, and other crafts gained renown for their artistic qualities as well as for their ingenious functionality.[73]

In addition, the communitarian aspects of tribal life were increasingly regarded not so much as the product of primitive peoples but as perhaps a more advanced state of social relations. For many intellectuals, Indians living in tightly-knit communities in close contact with nature appeared to be refreshing rejoinders to the dilemmas and disillusionment of modern, industrialized life. One activist deeply affected by what he perceived as an organic democracy practiced by the Pueblo Indians at Taos was John Collier. He envisioned Indian society as a "Red Atlantis" that held out vital clues for how to save modern society but that was gravely imperiled by uncaring policymakers and the impersonal forces of development. Collier became an unrelenting critic of the Indian bureau and a general nuisance to those who controlled Indian affairs.[74]

Many of the writings of members of the Five Tribes appealed to the sensibilities of white reformers. Essays, editorials, and longer works that focused on Indian history evoked an array of the finest human qualities displayed by tribal forebears, including honesty, loyalty, and bravery. Subtlely, tribal writers established a set of uniquely "Indian" characteristics; they culled the past for examples of advanced civilization and humanity, of men and women capable of profound artistic expression and environmental sensitivity. They dusted off and burnished the age old notion of the "Noble Savage."

Essays about religion underlined Indian spiritualism and mysticism, as did those that dealt with the environment. Indeed, ecological consciousness was often paired with Indians' spiritual reverence. Sunshine Rider, a Cherokee living in New York City, wrote an essay commending the Indians' harmonious relationship with nature, which was re-printed in *The American Indian*, a magazine published by the Five Tribes, in 1927. "The Redman," she pointed out, "unspoiled by the trappings of elaborate civilization, was responsive to all of Nature's moods. For him there were many things in the great wood that furnished his music and his logic." Rider's prose, growing increasingly purplish, brought Indians' love of nature into a state of religious ecstasy. "The Redman . . . recognized the Great Mystery in all creation, and believed that he drew from its spiritual power. He saw no need for only the seventh day of worship, since to him all days belonged to the Great Spirit of the Universe!"[75]

Perhaps nowhere did tribal writers turn prevailing ideas about Indians to their advantage as innovatively as when employing the racial calculus of the period. Just as surely as they portrayed the cultural achievements of their forebears, these authors founded their re-evaluations of the Indian past on the unique properties of biology. Tribal writers assented to a biological foundation of their distinctiveness, but in acknowledging the power of ancestry, they maintained that Indian blood was not a taint. Contrary to the prevailing wisdom of the day, they refuted the portrait of their ancestors as inherently inferior savages running wild through the forests. Rather, tribal writers depicted a noble race of peoples.

The ennoblement of Indian history had important ramifications. If Indian savagery was attributed to lineage, it followed that other traits might be similarly transmitted. If barbarity could be passed down, via biology, to later generations, it was also true that innate civilization could be handed down. The effort to depict Indians of the past as civilized, industrious, and inventive was thus an effort to bestow those qualities upon present-day Indians. Tribal writers argued that contemporary Indians were the heirs of the fine qualities of their forebears. They explicitly made this connection in their constant referral to blood. Tellingly, those behind *The American Indian* dedicated the magazine to "all those in whose veins flow the oldest aristocracy in the western hemisphere."[76]

Indian blood was potent. One letter writer in 1927 took issue with the way that Indians were depicted in a cartoon printed in a newspaper. Jack McCurtain used the supposed slight to declare that "time after time when the white man and Indian clashed wits, the Indian came out first."[77] He pointed to the battle at the Little Big Horn, in which Sitting Bull "out-generaled Custer and beat him at his own game," as one instance of Indian superiority and praised Indian athleticism as embodied by Jim Thorpe as another. So widely recognized were the exceptional racial traits of Indians that "[n]o less distinguished personage than the late ex-President Roosevelt expressed his regret that he did not have a strain of Indian ancestry. And surely so ambitious a man as he would not have desired to be of the Indian race if, in fact, that race is inferior."[78]

Blood contained the natural genius of Indian people. One need look no further than the lengthy list of great leaders who boasted the finest qualities of Indian civilization. In an article published in *Munsey's* magazine in 1914, John Oskison, the noted Cherokee writer, recounted a proud legacy of leadership. "From Dekanawida, who with Hiawatha founded the confederation of the five Iroquoian tribes in the fifteenth century, to Red Cloud, last survivor of the strong Sioux leaders, who died in 1909, the race has furnished real statesmen and strong war captains. . . . In the American Revolution, the names Red Jacket, Cornplanter, and Joseph Brant (Thayendanega) became known as those of a great orator, a great war chief, and a great peacemaker respectively."[79]

Indian leaders, tribal writers argued, helped give shape to colonial society. Oskison cited the seventeenth-century Pequot leader Metacom, "son of Massasoit," as a powerful chief who nearly succeeded in driving the English colonists in Virginia back into the ocean. "Because of his qualities of leadership he came to be known to the whites—and to later history—as King Philip."[80] Oskison implied that Indians produced men of genius and, for that matter, men who could slip comfortably into the cloak of royalty. Such leaders, he added, were in evidence throughout the continent, a fact that underscored the abilities of a range of Indian groups. He mentioned Osceola, "the Seminole leader in the war of 1835," and Tecumseh, who organized a pan-Indian movement in the early nineteenth century in an effort to defeat the Americans. Near the end of his list of prominent Indian figures, Oskison cited the peerless Sioux war chiefs, "Crazy Horse, Sitting Bull, Spotted Tail, Two Strikes, Little Crow, and the rest," who fought on long after others had succumbed.[81]

The greatness of Indian leaders, Oskison proposed, was transcendent, earning even the respect of hard-hearted whites. In his description of Osceola, Oskison highlighted not only the Seminole warrior's bravery and skill, but the disgrace of his death at the hands of his captors. He "died at the age of thirty-five—but not before a blaze of public indignation was kindled against his captors and his own name made an inspiration to the Indians who later fought to drive back a civilization they hated and feared." Similarly, Tecumseh, who picked up the mantle of King Philip and dreamed of a confederation of Indians in order to "sweep back the tide of white settlement which was drowning his people," came to represent profound qualities of bravery and acquired a status that was revered by later generations, white as well as Indian.[82]

Other tribal writers credited ancestry with bestowing superiority upon them. As a result, blood became contested territory. In one article published in *The American Indian*, a writer offered a celebration of Sequoyah, the creator of the Cherokee syllabry that made a written Cherokee language possible. Within the essay, the writer was obligated to dismiss claims that Sequoyah was descended from German parentage. Critical of the idea, the author of the essay maintained that one of the perpetrators of the lie "decided that no fullblood Indian could have had the brain and genius necessary for devising an alphabet."[83] Given Sequoyah's importance to the tribe—living proof, so to speak, of Indian creative intellect and brilliance—the threat that he might be "stolen" from the tribe and "awarded" to another people was too much. If Sequoyah could be taken from the Cherokees, they might be reduced to lesser beings of lesser ability.

The appeal to blood by tribal writers was well-suited to those white reformers with an affinity for Indian cultures. Unlike earlier "friends of the Indian," the reformers in the 1920s did not want to convert Indians into redhued versions of whites. They appreciated the unique flavor of Indian cultures and wondered if Indians should not be left alone to live as they pleased. Where Henry Dawes desired to bleach the Indian, to convert the "red savage" into an ordinary American, the new wave of reformers by the 1920s wished to retain elements of the native. Indeed, there was considerable worry that Indian peoples, and thus their cultures, were on the verge of extinction. Were Americans fated to live in a world devoid of the continent's first, and its most legitimate, inhabitant?[84]

Articles by tribal members provided portraits to address the concerns of white reformers as well as conform to the images they already possessed about Indian people. John Oskison's praise of past leaders fit the convention of the "good" Indian. The Indian war chief was a tragic figure. To be true to himself, he could not change his nature and become something else. In his article, Oskison reserved his highest regard for the imperious Sioux war chiefs who refused to meekly accede to white demands that they give up their old way of life in favor of being corralled on reservations. Only after whites launched a brutal campaign in which women and children became involved did the warriors give up their arms. Yet even in defeat, the luster of the "Indian spirit" was apparent.[85]

Despite the long contact between the tribes and whites, and the intermarriage that resulted, Indian blood was considered an agent that could not be stamped out. If clouded with other ancestral strains, it nonetheless exercised dominance. In 1926, a Choctaw writer and member of the Modern Language faculty at the University of Oklahoma sketched his autobiography in a piece published in *The American Indian*. Tracing the branches of his family tree, Todd Downing detailed a lineage not unusual for the region. Both grandfathers were white, one of whom traveled southward to Indian Territory from Indiana while the other migrated westward from Mississippi with his Choctaw-Chickasaw wife. The latter ancestor, owing to his marriage, involved himself with Choctaw affairs and in time acquired influence in tribal policymaking. Downing's father, in turn, similarly became "a power among the Choctaws." In their positions of influence, these men labored on behalf of the tribe and for its benefit. "At present," Downing wrote of his father, "he is a member of the Choctaw Tribal Council and is taking a leading part in the efforts to wrest from the United States government the fulfillment of promises which have never been fulfilled and to prevent further encroachment upon the rights of Indians."[86]

Yet despite the achievements of his family and his own experiences of living among the Choctaws, Downing staked his own claim to Indianness elsewhere. Instead of highlighting behavior that might mark him as Indian, he relied on something more basic. He ended his brief autobiographical account by stating: "Though only a quarter-blood Indian, that much gives [me] the proudest distinction [I] can ever gain—that of being a real American."[87]

To these views of blood there were, of course, dissenters. Some considered the dilution of Indian blood to have progressed too far. John Oskison, despite his unabashed glorification of past Indian civilizations, commented on the fate of vanishing Indians and maintained of the Five Tribes: "What a century of close association with whites has done to destroy the Indian identity of the Five Civilized Tribes, another half century will have accomplished with practically every tribe in the United States. Already the old type of Indian—the plains leader and camp-fire statesman, the mighty hunter and the poet-keeper of legends—has gone; and in two generations the Indian as a distinctive person will be all but a memory." Of his own tribe, he added, "there are not a hundred Cherokees in whose veins some white blood does not flow."[88]

Despite opposing views, however, many tribal members saw that blood was vital in keeping the tribes together. Even Oskison's comments suggest the primacy of blood. Indians had simply lost too much, he implied, to any longer be considered Indians. The only real Indians remaining, he stated, were now corralled on "inglorious" reservations. Even there, supposedly buffered from whites, "the corrupt soil of the reservation system has grown a new type of Indian—the one who intermarries with his white neighbor, who goes to school, raises cattle and horses, opens up and cultivates farms, learns a white man's trade and works at it."[89] Indianness, he implied, was biological.

Confronted with dangerous trends that undermined tribal attempts to protect their rights and resources, tribal writers saw the usefulness of themes that stressed Indian contributions to America. In the conceptions of some writers, Indians occupied a central role in American development. Indians were participants, not just extras, in all the great moments of American history. In one article, it was explained that the English colonists who came to the New World would not have survived had it not been for the kindness of the Indians who welcomed them ashore and subsequently shared their expansive knowledge of the North American environment.[90] In later eras, even as whites swallowed up huge chunks of real estate and broke their solemn treaties with various tribes, Indians responded with an eye toward fostering peace and compromise. In the face of such greed, one writer asserted, the Seminoles' "superiority [wa]s shown in the calm submission to the violation of treaties."[91]

In a portrayal of the Choctaw chief Pushmataha, the themes of Indian fairness and loyalty were explored further. In the early nineteenth century, Pushmataha was placed in a difficult position regarding white settlers moving westward and his fellow Indian leader Tecumseh. Having earlier given his word that he would not attack the whites, he nobly stood by that promise and refused Tecumseh's entreaties to join his pan-Indian alliance to snuff out white society. As Pushmataha allegedly stated, the decision whether or not to join with Tecumseh came down to "a question of carrying on that record of fidelity and justice for which our forefathers ever proudly stood."[92] Far from being savages or children who were unable to conceive of the value of statesmanship and living up to one's promises, the Choctaws were a people possessed of a strong moral compass and the will to enforce it. In a speech memorializing the great Choctaw chief, the Choctaw congressman, Charles David Carter, concluded that Pushmataha had "sav[ed] the white man's civilization west of the Alleghenies."[93]

One unmistakable objective in much of tribal writing from the period was the presentation of Indians as a proud people capable of highly advanced, civilized behavior. Indians were shown, time and again, to be different from, yet equal to, whites. Oskison's emphasis on Indian leaders made this quite clear. Indians exhibited special traits that helped them gain prominence. They made stunning contributions to American life and American history as a result. This core argument expressed understanding of a cultural pluralism that allowed Indians to remain distinct yet also be recognized for their contributions to American society. Sympathetic descriptions worked to mitigate criticisms and quiet the inferiority that whites so stubbornly attached to Indians.

* * *

The idea of a unified politics sparked significant tribal activity in the 1920s and 1930s. While generally portrayed as a period of almost complete disorganization and demoralization, tribal members mounted a multi-faceted response to their declining conditions. The grass-roots movements of conservatives, the more formal, organizational approach of progressives, and the subtle and innovative energies of a literary elite constituted a vigorous attempt to address the internal strife of the tribes in order to confront myriad outside threats.

Central to these efforts was the use of blood. Tribal members agreed that blood provided an essential marker of Indianness. Race could be used to counter notions of inferiority. Tribal members grounded their interpretations of the Indian past on conceptions of race. They embraced the fact of their biology: Indians had Indian blood. Indeed, they were Indians *because* they had Indian blood. Blood became the unifier that could preserve Indian distinctiveness in a hostile society. Reversing the general trend among whites of regarding Indian ancestry as a taint, tribal members firmly argued for their superiority on the basis of blood. In so doing, they helped to lay the groundwork for subsequent eras of Indian political development. The suggestion of a prideful racial heritage lingered long beyond the 1920s and 1930s. By the 1960s and 1970s, in fact, blood would prove important in political resurgence among native peoples. Although it was often used in the early twentieth century with ambiguity and inconsistency, blood mattered.

EIGHT
Epilogue

In 1934, Commissioner of Indian Affairs John Collier traveled to eastern Oklahoma to promote a new set of policies contained within the Indian Reorganization Act. Collier's "Red New Deal" promised to end what he viewed as the previous half-century's wrongheaded and destructive approaches to the "Indian problem." The IRA promised to return governance to the tribes and to halt the loss of Indian-owned lands. Tribes that assented would be eligible to draw up constitutions and begin the long journey back toward autonomy.[1]

Having hatched such a sweeping and radical plan, Collier was aggrieved upon arriving in Oklahoma to find his proposals drawing not praise from the various Indian groups across the state but skepticism. A peppering of criticisms followed his trail as he met with various tribes. Members of the Five Tribes, in particular, appeared less than thankful for Collier's "charitable" view of Indians and their ability to govern themselves. They complained that his plans would likely undo their many accomplishments and return them to a state of primitiveness.[2]

It did not escape notice that the tribal voices raining down upon Collier emanated from both progressive and conservative elements of the tribes. While internal turmoil and factionalism had not been banished from tribal affairs, a strong hint of unity was apparent. Although some tribesmen appreciated aspects of Collier's plan, including the offer to make funds available for community development, they were wary of the act in its entirety.[3] As a result of his disappointing trip, Collier was forced to allow Oklahoma's tribes an exemption from the IRA and subsequently to draft a separate bill to accommodate his critics. In 1936, the Oklahoma Indian Welfare Act was passed taking into account the demands of Indians residing in that state.[4]

The history of the Five Tribes since 1936 has not been harmonious. In fact, recent upheaval among the Cherokees in the winter and spring of 1998 brought about a constitutional crisis that threatened to bring down a tribal government that had made significant strides toward sovereignty over the course of several decades.[5] But while factionalism continues to plague tribal politics—as it does in many other polities in the country, Indian and non-Indian alike—it little resembles the period when tribal structures and commonly owned lands were dismantled, leaving tribal members to go their separate ways. Quite to the contrary, current dilemmas indicate that diverse tribal populations are engaged in real debates over real issues, a sign that their institutions are functioning and able.

In the difficult first decades of the twentieth century, members of the Five Tribes were bitterly divided. They lived in a stratified world where class, culture, and, increasingly, blood separated Indian from Indian. Indeed, the very notion of who was Indian and what that meant was subject to debate. But in the face of great hardship, tribal members mounted an effort to re-knit their populations. In so doing, they utilized the differences among them, their diverse sets of talents and values and beliefs, to fashion unity. Perhaps more than any other attribute, they used blood as the foundation for this daunting project. Revealing themselves to be adept manipulators of race in the tri-racial climate of Oklahoma, they employed blood both to afford a more inclusionary approach to membership and to ward off threat from outside.

John Collier encountered the handiwork of the Five Tribes. He met groups of tribal members who perhaps understood him better than he understood them. According to some scholars, Collier's method of activism, first honed among immigrants in New York City before finding new pastures among the Indians of Taos, NM, and elsewhere, was to offer the appearance of autonomy to a given minority group while retaining hold of real power himself.[6] In eastern Oklahoma, he found tribal members who knew where the real power lay and who had begun to speak with one voice in order to regain it.

Endnotes

CHAPTER ONE

1. Quoted in Muriel Hazel Wright, "A Chieftain's 'Farewell Letter' to the American People," in Daniel F. Littlefield, Jr., and James W. Parins, *Native American Writing in the Southeast: An Anthology, 1875–1935* (Jackson: University of Mississippi Press, 1995), 223–5.
2. Wright: 223.
3. Wright: 225.
4. Angie Debo, *The Road to Disappearance: A History of the Creek Indians* (Norman: University of Oklahoma Press, 1941). Also, Debo, *And Still the Waters Run* (Princeton: Princeton University Press, 1940).
5. These are but two, broad categories within the tribes, and there existed numerous smaller groupings and coalitions that added to the complexity of this factionalism. But for our purposes here, identifying "progressives" and "conservatives" gives a sense of just how far apart members within the same tribe could be. See Duane Champagne, *Social Order and Political Change: Constitutional Governments Among the Cherokee, the Choctaw, the Chickasaw, and the Creek* (Stanford: Stanford University Press, 1992).

CHAPTER TWO

1. Quoted in Angie Debo, *And Still the Waters Run* (Princeton: Princeton University Press, 1940), 13–15.
2. Quoted in Debo, *And Still the Waters Run*, 13–15.
3. Prior to the Civil War, the tribes had owned, as compensation for removing from their original homelands in the American Southeast, nearly the whole of present day Oklahoma. But after siding with the neighboring Confederacy in the Civil War, the tribes were forced in 1866 to sign new

treaties with the United States that effectively halved their domain. To the west, on land which the tribes had previously claimed for their exclusive use, the federal government planned to locate other displaced Indians and, in 1889, created Oklahoma Territory for white settlement. The Five Tribes were thus left with the eastern portion of their former territory, a still sizable estate of more than twenty million acres. See Debo, *And Still the Waters Run*; Danney Goble, *Progressive Oklahoma: The Making of a New Kind of State* (Norman: University of Oklahoma Press, 1980).

4. Ralph E. Olson, "Agriculture in Oklahoma," in John W. Morris, ed., *Geography of Oklahoma* (Oklahoma City: Oklahoma Historical Society, 1977), 69.

5. Debo, *And Still the Waters Run*, 13; Grant Foreman, *The Five Civilized Tribes* (Norman: University of Oklahoma Press, 1934); Arrell M. Gibson, *The History of Oklahoma* (Norman: University of Oklahoma Press, 1984).

6. Duane Champagne, *Social Order and Political Change: Constitutional Governments Among the Cherokee, the Choctaw, The Chickasaw, and The Creek* (Stanford: Stanford University Press, 1992), 212.

7. All historians of the Five Tribes use some or all of these labels, especially those referring to blood. Angie Debo, the principal documenter of the tribes during the early twentieth century, applied blood liberally in her writings, spanning from the 1930s to the 1970s. In contrast, Duane Champagne does not use mixed or full blood at all in his dense study of tribal institutions. Some writers have struggled in print with the terms, as pertained to the Five Tribes and to other tribes. See "Note about terms" at the beginning of this study.

8. Duane Champagne, *Social Order and Political Change*, 212.

9. Champagne, 212.

10. Quoted in Craig Miner, *The Corporation and the Indian: Tribal Sovereignty and Industrial Civilization in Indian Territory, 1865–1907* (Columbia: University of Missouri Press, 1976), 119.

11. On education see Daniel F. Littlefield, Jr., *Alex Posey: Creek Poet, Journalist, and Humorist* (Lincoln: University of Nebraska Press, 1992); Champagne, *Social Order and Political Change*; Devon A. Mihesuah, *Cultivating the Rosebuds: The Education of Women at the Cherokee Female Seminary, 1851–1909* (Urbana-Champagne: University of Illinois Press, 1993); K. Tsianina K. Lomawaima, *They Called it Prairie Light: The Story of Chilocco Indian School* (Lincoln: University of Nebraska Press, 1994).

12. Miner, *The Corporation and the Indian*, Chapter Six.

13. Debo, *And Still the Waters Run*, 15–16.

14. The influence of progressive notions of private property ownership were ascendant throughout the nineteenth century. For a discussion of how mixed blood segments of the Choctaws used tribal law to promote their own interests and the development of private property in the early nineteenth century, see Clara Sue Kidwell, *Choctaws and Missionaries in Mississippi, 1818–1918* (Norman: University of Oklahoma Press, 1995).

Endnotes

15. Miner, *The Corporation and the Indian*; Goble, *Progressive Oklahoma*, 62-5; Jeffrey Burton, *Indian Territory and the United States, 1866-1906: Courts, Government, and the Movement for Oklahoma Statehood* (Norman: University of Oklahoma Press, 1995), 106-10.

16. The Chickasaws elected a "governor" rather than "chief." Arrell M. Gibson, *The Chickasaws* (Norman: University of Oklahoma Press, 1971).

17. The impulse to centralize tribal government had begun early in the nineteenth century when the tribes still resided in the Southeast. The Cherokees were the first to draft a constitution in the 1820's, and in 186, the Creeks became the last of the tribes to formally place power in a central government and ratify a constitution. Previously, power had been decentralized and revolved around a number of tribal towns. Towns had been, and to a degree continued to be in the West, the center of tribal members' lives. Local leaders had exercised influence while participating in larger networks by which allied towns coordinated preparations for war or negotiations for peace. Towns were also the centers of religious and social activity, and local religious leaders exercised profound influence. Champagne, *Social Order and Political Change*, 13-49; Debo, *And Still the Waters Run*, 18, and *The Road to Disappearance: A History of the Creek Indians* (Norman: University of Oklahoma Press, 1941), 9-12.

18. Champagne, *Social Order and Political Change*, 124-75.

19. Among the myriad studies of the Removal of the Five Tribes, see Grant Forman, *Indian Removal: The Emigration of the Five Civilized Tribes of Indians* (Norman: University of Oklahoma Press, 1972). For the re-establishment of tribal order in Indian Territory, see, for example, William McLoughlin, *After the Trail of Tears: The Cherokees' Struggle for Sovereignty, 1839-1880* (Chapel Hill: University of North Carolina Press, 1993).

20. For studies of the Five Tribes and the Civil War, see Annie H. Abel, *The American Indian and the End of the Confederacy, 1863-1866* (Cleveland: Arthur H. Clark, 1925); Laurence M. Hauptman, *Between Two Fires: American Indians in the Civil War* (New York: Free Press, 1995).

21. Champagne, *Social Order and Political Change*, 220-3.

22. Miner, *The Corporation and the Indian*; Goble, *Progressive Oklahoma*, 57-62.

23. Champagne, *Social Order and Political Change*, 215.

24. Ibid.

25. Ibid.

26. Goble, *Progressive Oklahoma*, 57; also, for a good discussion of the reverse social status of whites and tribal members in late nineteenth-century Indian Territory, see the introduction to Daniel F. Littlefield, Jr., *Seminole Burning: A Story of Racial Vengeance* (Jackson: University of Mississippi Press, 1996), 23.

27. Littlefield, *Seminole Burning*, 23.

28. Champagne, *Social Order and Political Change*, 211.

29. Goble, *Progressive Oklahoma*, 46–51; Murray R. Wickett, "Contested Territory: Whites, Native Americans, and African-Americans in Oklahoma, 1865–1907," (unpublished Ph.D. dissertation, University of Toronto, 1996), 115–17; Debo, *And Still the Waters Run*, 11–14.

30. Katja May, *African Americans and Native Americans in the Creek and Cherokee Nations, 1830s to 1920s: Collision and Collusion* (New York: Garland Publishing, 1996), 69.

31. See Debo, *And Still the Waters Run*, 11–14; Champagne, *Social Order and Political Change*, 210–13.

32. Champagne, *Social Order and Political Change*, 233–4.

33. Debo, *And Still the Waters Run*, 11.

34. Champagne, *Social Order and Political Change*, 210–12.

35. Debo, *And Still the Waters Run*, 11; see also Daniel F. Littlefield, Jr., *The Chickasaw Freedmen: A People Without a Country* (Westport: Greenwood Press, 1980).

36. Quoted in Debo, *And Still the Waters Run*, 11.

37. Littlefield, *Chickasaw Freedmen*, 45–46; Gibson, *The Chickasaws*; see also Burton who cites figures as of 1890: 3,800 blacks and 3,100 Chickasaws "by blood," *Indian Territory and the United States*, 180.

38. Debo, *The Road to Disappearance*, 268–81; Katja H. May, "Collision and Collusion: Native Americans and African Americans in the Cherokee and Creek Nations, 1830s to 1920s," (unpublished Ph.D. dissertation, University of California at Berkeley, 1994), 194–209; Kenneth W. McIntosh, "Chitto Harjo, the Crazy Snakes and the Birth of Indian Political Activism in the Twentieth Century," (unpublished Ph.D. dissertation, Texas Christian University, 1993), 24–5.

39. Champagne, *Social Order and Political Change*, 220–21.

40. Ibid., 223–37.

41. For tribes' exemption from Dawes Act, see Francis Paul Prucha, *The Great Father: The United States Government and the American Indians*, vol. II, Lincoln, 1984.

42. For allotment, see Prucha, *The Great Father*, Chapters 26, 29, 35; Frederick E. Hoxie, *A Final Promise: The Campaign to Assimilate the Indians, 1880–1920* (Lincoln: University of Nebraska Press), 1984, Chapter 5; Donald L. Parman, *Indians and the American West in the Twentieth Century*, Bloomington, 1994, Chapters 1, 3.

43. Parman, *Indians and the American West*, 8.

44. Land rushes in what was known as the "Cherokee Strip," land that previously belonged to the Cherokees but was ceded in the 1866 treaties, alerted the tribes of coming threats. By allowing whites to settle to the tribes' immediate west, the land rushes made the Five Tribes the inhabitants of an island surrounded on all sides by an increasingly demanding white society. See Goble, *Progressive Oklahoma*.

45. For statehood, see Goble, *Progressive Oklahoma*, Chapter Nine.

Endnotes

46. The termination dates for the tribes varied according to their agreements with the federal government.

47. Quoted in Debo, *And Still the Waters Run*, 132.

CHAPTER THREE

1. Quoted in Debo, *And Still the Waters Run* (Princeton: Princeton University Press, 1940), 59. See also Daniel F. Littlefield, Jr., "Utopian dreams of the Cherokee Fullbloods, 1890–1930," *Journal of the West* 10 (Spring, 1971).

2. Quoted in Debo, *And Still the Waters Run*, 59–60.

3. Debo, 47. Some tribes showed greater disparity than others; among the Chickasaws, of a total population of 6,319, a mere 1,538, or twenty four percent, were full bloods. The Creeks, Seminoles, and a group of Mississippi Choctaw, however, contained more full bloods than mixed bloods. Also, Arrell M. Gibson, *The Chickasaws* (Norman: University of Oklahoma Press, 1971), 274.

4. Debo, *And Still the Waters Run*, 98–101.

5. Harjo was also known by the English name Wilson Jones. See Kenneth W. McIntosh, "Chitto Harjo, the Crazy Snakes and the Birth of Indian Political Activism in the Twentieth Century," (unpublished Ph.D. dissertation, Texas Christian University, 1993), iii.

6. Daniel F. Littlefield, Jr., *Alex Posey: Creek Poet, Journalist, and Humorist* (Lincoln: University of Nebraska Press, 1992), 143–6.

7. McIntosh, "Chitto Harjo," 48–9; Littlefield, *Alex Posey*, 143.

8. Quoted in Littlefield, *Alex Posey*, 197.

9. McIntosh, "Chitto Harjo," 52.

10. Georgia Rae Leeds, *The United Keetoowah Band of Cherokee Indians in Oklahoma* (New York: Peter Lang, 1996), 7–8.

11. Leeds, *The United Keetoowah Band*, 8; William G. McLoughlin, "Syncretism: The Origins of the Keetoowah Society, 1854–1861," in McLoughlin, Walter H. Conser, ed., *The Cherokees and Christianity, 1794–1870: Essays on Acculturation and Cultural Persistence* (Athens: University of Georgia Press, 1994); Debo, *And Still the Waters Run*, 54.

12. Leeds, *The United Keetoowah Band*, 8.

13. Ibid., 10.

14. For Keetoowah, and Cherokee, values and communitarian principles, see McLoughlin, "Syncretism."

15. Champagne discusses the clash of values and economic systems in *Social Order and Political Change*, 214–16.

16. The widespread looting of tribal assets, which is discussed further in Chapters Five and Six, have been most exhaustively and grippingly recounted by Debo in *And Still the Waters Run*.

17. Leeds, *The United Keetoowah Band*, 9.

18. McIntosh, "Chitto Harjo," 52.

19. Debo, *And Still the Waters Run*, 53

20. Littlefield, *Alex Posey*, 146.

21. This claim is surely inflated, and it might represent the total number of full bloods within the tribes. See Debo, *And Still the Waters Run*, 54; McIntosh, "Chitto Harjo," 167. While the exact date of the group's founding is unknown, Debo places it at 1895. McIntosh maintains that its existence remained a secret from federal authorities until 1906.

22. McIntosh, "Chitto Harjo," Chapter Five.

23. See Jack D. Forbes, *Black Africans and Native Americans: Color, Race, and Caste in the Evolution of Black-Red Peoples*. (Oxford: Basil Blackwell, 1988).

24. Reginald Horsman, "Scientific Racism and the American Indian in the Mid-Nineteenth Century," *American Quarterly* 27 (May, 1975).

25. Nancy Shoemaker, "How Indians Got to Be Red," *American Historical Review* (June, 1997), 625–644.

26. McLoughlin notes that the debate of separate species was then swirling around the South, where various white thinkers were engaged in a project to define racial origins and justify slavery. Chapter 6, "Christianity and Racism: Cherokee Responses to the Debate over Indian Origins, 1760–1860," in McLoughlin, Conser, ed., *The Cherokees and Christianity*. Also, McLoughlin writes that some tribal members used tribal origin stories to prove their racial difference from other groups. The lesson of tribal myths, he writes, was that "there is a fundamental separation between the red and white people. *Cherokee Renascence in the New Republic* (Princeton: Princeton University Press, 1986), 177.

27. Burton, *Indian Territory and the United States*, 180.

28. Katja May, *African Americans and Native Americans in the Creek and Cherokee Nations, 1830s to 1920s: Collision and Collusion* (New York: Garland Publishing, 1996), 69.

29. There were times, though, when blood drew firm lines between tribal members, as during the Civil War. In the instance of the Cherokees, the Keetoowah Society renounced efforts by the tribal government to ally with the Confederacy and explicitly stated that its members were restricted to full bloods. See McLoughlin, "Syncretism."

30. There had been instances of racial division prior to the end of the nineteenth century, but these had proved more flexible; they often accommodated for individuals who did not possess the racial makeup that a group claimed. In the 1850s, for example, the Keetoowah Society reacted to the coming danger of the Civil War by criticizing those Cherokees who exhibited pro-southern sympathies. These tribal members, the Keetoowahs stated, were sacrificing their loyalty to the tribe for their own personal interests. They were demonstrating improper behavior, behavior which marked them as being more white than Indian. Above all, Cherokee supporters of the Confederacy, who were largely mixed blood, forfeited their rights as Cherokees. In contrast, the Keetoowahs required its members to be full bloods. This requirement, however, was frequently overlooked and the group associated itself with white Baptists ministers in the territory who were out-

Endnotes

spoken in their criticism of slavery and the Confederacy. McLoughlin, "Syncretism."

31. Quoted in Littlefield, *Alex Posey*, 197.
32. Ibid.
33. Typescript of an editorial in *The Tulsa Democrat*, October 4, 1901, "Editorial on John McIntosh, et. al." In Cherokee Nation Papers, Box 176, Folder 007616, Western History Collection, University of Oklahoma (WHC).
34. Wilson Jones to Sec. of Interior, December 4, 1900, R. S. Cate Collection, Box 13, Folder 6 (Timothy Johnson Papers), WHC.
35. Quoted in Debo, *And Still the Waters Run*, 154.
36. See Reginald Horsman, "Scientific Racism and the American Indian in the Mid-Nineteenth Century," in *American Quarterly* 27 (May, 1975), 152–168.
37. *Muskogee Times Democrat*, December 30, 1904.
38. Debo, *And Still the Waters Run*, 131–2.
39. Debo, 49.
40. Debo, 57.
41. Quoted in Littlefield, *Alex Posey*, 202.
42. Quoted in Littlefield, 202.
43. There is a rich literature on the issue of Indian stereotypes and racial thinking. See, among others, Richard Berkhofer, *The White Man's Indian: Images of the American Indian from Columbus to the Present* (New York: Alfred A Knopf, Inc., 1978); Frederick E. Hoxie, *A Final Promise: The Campaign to Assimilate the Indian, 1880–1920* (Lincoln: University of Nebraska Press, 1984).
44. There are even recorded visits by tribal members to Mexico. Littlefield, *Alex Posey*, 241–242; Littlefield, "Utopian Dreams of the Cherokee Fullbloods": 413, 418–19.
45. The Creek Snakes wanted to reestablish their "old government in Mexico," Littlefield, *Alex Posey*, 197–202. Also see Littlefield, "Utopian Dreams of the Cherokee Full Bloods." Record Group 75, at the National Archives and Federal Archives and Records Center (FARC) Southwest Region, contain various instances of emigration schemes. Various freedmen of the Five Tribes also considered emigration as a way to escape persecution by whites. Some tried to go north, to Canada, but there they met with considerable hostility. Bateman, "'We're Still Here," 175–9.
46. Quoted in Debo, *And Still the Waters Run*, 59–60.
47. Thus the title of Littlefield's, "Utopian Dreams of the Cherokee Freedmen."
48. Eufaula Harjo began the Indian Bureau after growing dissatisfied with the Creek Snakes. Eufaula Harjo to Francis Leupp, July 30, 1909, BIA CCF, Creek, RG 75, FARC, Southwest Branch, Fort Worth, Texas.

CHAPTER FOUR

1. Additional symbols of the joining of the races were also on display: the name of the new state was Choctaw for "land of the red people," and its official seal incorporated images representing the Five Civilized Tribes as well as a rendering of an Indian and a frontiersman shaking hands in friendship. Angie Debo, *And Still the Waters Run* (Princeton: Princeton University Press, 1940), 171, 212.

2. The contradictory views of white Americans toward Indians included, and continue to include, the notion that white ancestry could provide a leavening attribute when mixed with Indian ancestry. In contrast to views of African Americans, for whom the one-drop rule is rigidly enforced and who are therefore excluded from the whitening process of racial intermingling, Indians occupy a different strata of racial hierarchies. There were, however, different types of Indians, and whites considered the prospects of full blood advancement less likely than those of mixed bloods. Further discussion below. See, among others, Walter B. Michaels, *Our America: Nativism, Modernism, and Pluralism* (Durham: Duke University Press, 1994); Daniel F. Littlefield, Jr., *Alex Posey: Creek Poet, Journalist, and Humorist* (Lincoln: University of Nebraska Press, 1992); Murray R. Wickett, "Contested Territory: Whites, Native Americans, and African-Americans in Oklahoma, 1865–1907," (unpublished Ph.D. dissertation, University of Toronto, 1996).

3. For allotment in Indian Territory, see Francis P. Prucha, *The Great Father: The United States Government and the American Indians*, Vol. II (Lincoln: University of Nebraska Press, 1984); Donald L. Parman, *Indians and the American West in the Twentieth Century* (Bloomington: Indiana University Press, 1994); Janet McDonnell, *The Dispossession of the American Indian, 1887–1934* (Bloomington: Indiana University Press, 1991).

4. The increasingly complex matter of determining competency among Indians became an important part of implementing allotment and subsequent legislation regarding Indian resources. Janet McDonnell, "Competency Commissions and Indian Land Policy, 1913–1920," *South Dakota History* 11 (Winter, 1980): 21–34.

5. Parman, *Indians and the American West*, 53; Prucha, *The Great Father*, Chapter 34; Debo, *And Still the Water Runs*, 281.

6. Prucha, *The Great Father*, Chapter 34.

7. Owen, who had controlled huge parcels of Cherokee country, as much as 40,000 acres, in the late nineteenth century, arranged to lease and buy, at exceptionally favorable rates, much of that acreage in the aftermath of Oklahoma statehood. His personal interests in the matters of restrictions and federal oversight of any transactions involving land, thus, were great. Nonetheless, his views were echoed by other tribal progressives who desired the freedom to enter the market unencumbered and unpenalized by the federal authorities. Debo, *And Still the Waters Run*, 98–9.

8. Debo, *And Still the Waters Run*, 175.

9. Owen, it should be noted, had an ulterior motive in his campaign to lift restrictions on progressives. He and other prominent men desired to reduce what they considered the meddlesome federal presence in the region in advance of statehood. Owen was among the signers at a meeting in Okmulgee in 1904 of a resolution requesting the transfer of tribal affairs from the Department of the Interior to local authorities. The resolution's signers stated that "[i]t is the sincere conviction of this convention that the best thing that can be done for even our most helpless Indian fellow-citizens is to place their protection with the courts, which normally protects the interests of the helpless and incompetent, and in all respects to place them as rapidly as possible in the position of ordinary citizens, so that instead of being American Indians, these people who surely have the fullest rights to the title, may become in name and fact American citizens." Subsequently, at the Trans-Mississipi Commercial Congress at Muskogee in 1907, he reiterated his beliefs in response to a speech by the Creek chief Moty Tiger, who maintained that Indian allotments required stiffer protections. Owen maintained that the interests of Indians were best served not by the federal government, but rather by the "generous hearted sons of Oklahoma." "They are our children," he said about the restricted full bloods, "and we want to take care of our own children and we don't want any stepmother." Yet, it should be noted, even as he tried to manuever control of Indian affairs away from the federal government and into the hands of the state, Owen nonetheless agreed with the inclination among federal officials to distinguish between tribal members. Debo, *And Still the Waters Run*, 141, 173–5.

10. The rolls are described in Littlefield, *Alex Posey*; Debo, *And Still the Waters Run*; Goble, *Progressive Oklahoma: The Making of a New Kind of State* (Norman: University of Oklahoma Press, 1980); Parman, *Indians and the American West*, 52–3. Russell Thornton recounts that the Cherokees during much of the nineteenth century used lineage to determine tribal membership. The Dawes Rolls, subsequently, became the essential informationo base for determining tribal membership. Thornton, *The Cherokees: A Population History* (Lincoln: University of Nebraska Press, 1990), 140.

11. Indeed, many mixed bloods, who were often encouraged to apply for competency by land sharpers, were quite unable to take charge of their own affairs and soon lost their lands and their livelihoods. Conversely, a number of full bloods who possessed the capacity and desire to enter the land market were denied the opportunity.

12. D. Beaulieu, "Curly Hair and Big Feet: Physical Anthropology and the Implementation of Land Allotment on the White Earth Chippewa Reservation," *American Indian Quarterly* 8 (Fall, 1984): 281–314; Melissa Meyer, *The White Earth Tragedy: Ethnicity and Dispossession at a Minnesota Anishinaabe Reservation* (Lincoln: University of Nebraska Press, 1994).

13. Wickett, "Contested Territory," 42, 353.

14. Hoxie, *A Final Promise: The Campaign to Assimilate the Indian, 1880–1920* (Lincoln: University of Nebraska Press, 1984), 181–2; Prucha, *The Great Father*, 882–3. This topic is discussed further in Chapter Five.

15. Littlefield, *Alex Posey*, 245.

16. W. David Baird, ed., *A Creek Warrior for the Confederacy: The Autobiography of Chief G. W. Grayson* (Norman: University of Oklahoma Press, 1988).

17. Wickett, "Contested Territory," 85; Katja May, "Collision and Collusion: Native Americans and African Amerians in the Cherokee and Creek Nations, 1830s to 1920s," (unpublished Ph.D. dissertation, University of California at Berkeley, 1994), 99.

18. On attitudes of elite tribal members toward white settlers, Daniel F. Littlefield, Jr., *Seminole Burning: A Story of Racial Vengeance* (Jackson: University of Mississippi Press, 1996).

19. Wickett, "Contested Territory," 85.

20. Ibid., 89.

21. Ibid., 52.

22. Kenneth W. McIntosh, "Chitto Harjo, The Crazy Snakes and the Birth of Indian Political Activism in the Twentieth Century," (unpublished Ph.D. dissertation, Texas Christian University, 1993), 83.

23. Daniel F. Littlefield, Jr., and Lonnie E. Underhill, "The 'Crazy Snake Uprising' of 1909: A Red, Black, or White Affair?" *Arizona and the West* XX (Winter, 1978): 307–324; McIntosh, "Chitto Harjo," 118–127.

24. Norman L. Crockett, *The Black Towns* (Lawrence: University of Kansas Press, 1979); Jimmie L. Franklin, *Journey Toward Hope: A history of Blacks in Oklahoma*, Norman, 1982; Arthur L. Tolson, *The Black Oklahomans: A History, 1541–1972* (New Orleans: Edwards Printing Co., 1972); Goble, *Progressive Oklahoma*; May, *African Americans and Native Americans in the Cherokee and Creek Nations*.

25. Rebecca B. Bateman,, "We're Still Here': History, Kinship, and Group Identity Among the Seminole Freedmen of Oklahoma," (unpublished Ph.D. dissertation, Johns Hopkins University, 1991), Chapter Five; Wickett, "Contested Territory," Chapter Six.

26. For example, in September 1911, the *Tahlequah Arrow* reported a "race war" "raging" in Caddo, a town in Choctaw and Chickasaw country, where blacks and whites exchanged gunfire. After a white man was killed, blacks were forced to flee the town. "For the first time in its history, Caddo Sunday night had not a single negro resident, the blacks having all fled from much of the surrounding territory. The blacks took fright at the temper of the whites and feared to remain another night." The reporter noted, however, that the "exodus" was carried out in a somewhat disciplined manner, and railroad officials dispatched "extra cars" for the refugees. The troubles in Caddo did not appear to come as a surprise, and people did not flee wildly into the night. "All outgoing trains were crowded," the newspaper reported, "while extra cars were required for the handling of the baggage and express."

Endnotes 121

When all preparations were complete, "[a] large crowd assembled at the depot [and] cheered each departing train which carried the blacks from the town." *Tahlequah Arrow*, September 7, 1911. In Coweta, a town located between Muskogee and Tulsa in Creek country, the militia was sent in to calm tensions after racial violence in 1911 left a number of people dead. According to an observerer, the scene was grisly: "D.J. Beavers, city attorney of Coweta, Ok., and a graduate of Leland Stanford University, was shot through the head and instantly killed; Stellar Thompson and Carman Oliver, two white men were shot through the body and may die, Ed Suddeth, negro, was shot, arrested, strung up to a water tank, cut down, jailed, started out of town by officers, and shot to death by a mob, while Ed. Ruse another black who caused all the trouble was shot in the leg and jailed in a race riot which broke out here Sunday afternoon, growing out of the negro pushing a white woman off the sidewalk." *Tahlequah Arrow*, October 26, 1911.

27. Wickett, "Contested Territory," 376; McIntosh, "Chitto Harjo," 114.
28. *Tahlequah Arrow*, July 21, 1910.
29. May, "Collision and Collusion," 279; Daniel F. Littlefield, Jr., *The Cherokee Freedmen: From Emancipation to American Citizenship* (Westport: Greenwood Press, 1978).
30. Thornton, *The Cherokees*, 140.
31. Quoted in Wickett, "Contested Territory," 64.
32. Quoted in Littlefield, *The Chickasaw Freedmen: A People Without a Country* (Westport: Greenwood Press, 1980), 66.
33. Ibid., 66.
34. May, "Collision and Collusion," 101. This was not the earliest miscegentation law passed by the tribe.
35. Wickett, "Contested Territory," 350–1.
36. Bateman, "'We're Still Here,'" 161–2.
37. Ibid., 161–2.
38. The term is Bateman's, Ibid., 253.
39. May, *Native Americans and African Americans*, 178, 199. Intermarriage with whites, on the other hand, was relatively common.
40. Wickett, "Contested Territory," 80.
41. Quoted in Ibid, 411–12.
42. Wickett, 412.
43. Littlefield, *The Chickasaw Freedmen*, 208–10, 216.
44. Quoted in Debo, *And Still the Waters Run*, 135. See also McIntosh, "Chitto Harjo," and Wickett, "Contested Territory."
45. Quoted in Debo, *And Still the Waters Run*, 128.
46. Scholars investigating the racial interplay of the statehood period generally characterize the anti-black attitudes of tribal members as passive responses to the actions of whites, this explanation is a bit simplistic. Interestingly, two recent studies of Indian-black relations in eastern Oklahoma both employ the same language to describe the deterioration of relations between tribal members and freedmen. Katja May notes that as a

result of the Dawes Rolls' classifications, "Indian identity had thus become determined by European Americans." In turn, "[a] wedge was being driven separating Native Americans from African Americans who had formerly shared culture and history." But she also notes that while the Creeks were more disposed toward friendship with blacks, blacks felt the need to establish separate towns within Creek country. Her scant findings of intermarriage between the races further argues against her point. Rebecca Bateman, in her study of Seminole freedmen, credits statehood, and the attendant legal designation of Indians as whites, with "[driving] a widening wedge between Indians and blacks within the Seminole tribe." While the policies of the United States government and the actions of white settlers undoubtedly influenced tribal attitudes toward blacks, the souring of race relations must also be seen to stem from the independent determinations of the tribes. Assessing their situation in Oklahoma during a period of great change, tribal members positioned themselves in the most favorable light possible. They were not passive pawns being manipulated by white Americans. And while the federal government and white society might have deepened the divide between Indans and blacks, the "wedge" had long existed. May, "Collision and Collusion," 367; Bateman, "'We're Still Here,'" 162. McIntosh notes that many Snakes at Hickory Ground wanted the blacks who had encamped there to leave, "Chitto Harjo," 21–2.

47. "Will Swat the Negro," *Tahlequah Arrow,* July 21, 1910, re-printed from the *Muskogee Times-Democrat.*

48. Littlefield, *Alex Posey,* 8–9.

49. *Muskogee Times-Democrat,* December 7, 1906.

50. Wickett, "Contested Territory," 405, 407.

51. Quoted in Ibid., "Contested Territory," 412.

52. Quoted in Ibid., "Contested Territory," 412.

CHAPTER FIVE

1. Susie Ellis to Gabe E. Parker, October 1917, Bureau of Indian Affairs Central Classified Files, Cherokee Nation, Decimal file 723, National Archives, Washington, DC. (Hereafter referred to as BIA CCF, RG 75, NA).

2. Susie Ellis to Gabe E. Parker, October 1917, BIA CCF, RG 75, NA.

3. Angie Debo provides an overview of tribal miseries in, *And Still the Waters Run* (Princeton: Princeton University Press, 1940). Also, Danney Goble, *Progressive Oklahoma: The Making of a New Kind of State* (Norman: University of Oklahoma Press, 1980). The field matron Susie Ellis suggested that the government needed to provide assistance, as the tribal members "believe in the government." Ellis to Parker, October 1917, BIA CCF, Cherokee Nation, file 700, RG 75, NA.

4. E. W. Loomis to T. Roosevelt, October 31, 1907, BIA CCF, Cherokee Nation, file 723, RG 75, NA.

Endnotes

5. D. Spense to Dana H. Kelsey, U. S. Indian Superintendent, October 15, 1913, BIA CCF, Records of Supervising Field Clerks, Box 2, Night Hawks folder, RG 75, Federal Archives and Record Center, Southwest Region, Fort Worth, Texas. (Hereafter referred to as FARC Southwest.)

6. Spense to Kelsey, October 15, 1913, BIA CCF, Records of Supervising Field Clerks, Box 2, Night Hawks folder, RG 75, FARC Southwest.

7. G. W. Clark to Cato Sells, BIA Commissioner, March 10, 1917, BIA CCF, Cherokee Nation, file 723, RG 75, NA.

8. G. W. Clark to Cato Sells, BIA Commissioner, March 10, 1917, BIA CCF, Cherokee Nation, file 723, RG 75, NA.

9. Ben F. Smith to O. K. Chandler, January 11, 1917, BIA CCF, Cherokee Nation, file 723, RG 75, NA.

10. See Debo, *And Still the Waters Run*; Grant Foreman, *History of the Five Civilized Tribes* (Norman: University of Oklahoma Press, 1934); Arrell M. Gibson, *A History of Five Centuries* (Norman: University of Oklahoma Press, 1981).

11. Throughout the early statehood period, conservative tribal members continued to evade competency commissions and formal enrollment. They frequently refused their allotment certificates and tried to hang on to former lands. Many of these instances are recorded in the BIA Central Classified Files of Record Group 75 located both at the National Archives and at FARC Southwest. Also, Goble, *Progressive Oklahoma*, 72-6; Debo, *And Still the Waters Run*, 54-5.

12. In contrast, the northern and western regions of the state were settled by transplanted midwesterners who established larger farms and ranches. The distinctiveness of the immigrant streams leading into Oklahoma were clearly evident and have been commented upon by several scholars. See especially John Thompson, *The Closing of the Frontier: Radical Response in Oklahoma, 1889-1923* (Norman: University of Oklahoma Press, 1986); Howard R. Lamar, "The Creation of Oklahoma: New Meanings for the Oklahoma Land Run," in *The Culture of Oklahoma*, eds. Howard F. Stein and Robert F. Hill (Norman: University of Oklahoma Press, 1993).

13. Douglas Hale "The People of Oklahoma: Economics and Social Change," in Anne H. and H. Wayne Morgan, eds., *Oklahoma: New Views of the Forty-Sixth State* (Norman: University of Oklahoma Press, 1982), 44-7, 56-7.

14. The dominant cash crop soon became cotton, and Oklahoma farmers busily set the stage for a continuing rash of class conflict and soil depletion that would culminate in the dust bowl of the 1930's. See Garin Burbank, *When Farmers Voted Red: The Gospel of Socialism in the Oklahoma Countryside, 1910-1924* (Westport: Greenwood Press, 1976); James R. Green, *Grass-Roots Socialism: Radical Movements in the Southwest, 1895-1943* (Baton Rouge: Louisiana State University Press, 1978); Sheila Manes, "Pioneers and Survivors: Oklahoma's Landless Farmers," in Morgan, eds., *Oklahoma*.

15. Hale, "The People of Oklahoma," 44.

16. Ibid., 60.

17. Hale: 58. See also Manes, "Pioneers and Survivors," 102–10; Burbank, *When Farmers Voted Red*; Green, *Grass-Roots Socialism*.

18. Manes, "Pioneers and Survivors," 110–12; Green, *Grass-Roots Socialism*; Burbank, *When Farmers Voted Red*.

19. Hale, "The People of Oklahoma," 47.

20. Kenny L. Brown, "Progressivism in Oklahoma Politics, 1900–1913: A Reinterpretation, in Davis. D. Joyce, *'An Oklahoma I had Never Seen Before: Alternative Views of Oklahoma* (Norman: University of Oklahoma Press, 1994); Danney Goble, *Progressive Oklahoma: The Making of a New Kind of State* (Norman: University of Oklahoma Press, 1980). For usury, see Thompson, *Closing the Frontier*. For federal policy, see Francis Paul Prucha, *The Great Father: The United States Government and the American Indians*, vol II (Lincoln: University of Nebraska Press, 1984).

21. Realizing the need to stimulate agricultural production, the government launched a "Food and Feed" campaign. This program was designed to put idle fields under the plow in order to provide for the needs of foreign populations and armies. The lands of the Five Tribes were an attractive prize for the bureau. Various documents relating to the agricultural drive exist in BIA CCF RG 75, both in the National Archives and at FARC Southwest. See also David Wood, "American Indian Farmland and the Great War," *Agricultural History* LV (1981), 249–51; Janet A. McDonnell, *The Dispossession of the American Indian, 1887–1934* (Bloomington: Indiana University Press, 1981).

22. William Neighbors to Secretary of the Interior, April 15, 1917, BIA CCF, Five Tribes, File 910, RG 75, NA.

23. B. V. Hampton to C. D. Carter, April 19, 1917, BIA CCF, Five Tribes, File 910, RG 75, NA.

24. See McDonnell, *The Dispossession of the American Indian*; Prucha, *The Great Father*. For an overview of agricultural policies regarding Indians, see Richard Lewis, *Neither Wolf Nor Dog: American Indians, Environment, and Agrarian Change* (New York: Oxford University Press, 1994).

25. Debo, *And Still the Waters Run*, Chapter 6.

26. Ibid., 196–7.

27. Ibid., 117.

28. Goble, *Progressive Oklahoma*, 76; Debo, *And Still the Waters Run*.

29. Various reports in the BIA files explore Indian murders. Also, Debo, *And Still the Waters Run*.

30. Peter M. Wright, "John Collier and the Oklahoma Indian Welfare Act of 1936," *Chronicles of Oklahoma* L (1972), 352. Debo, *And Still the Waters Run*, 115–17.

31. Debo, *And Still the Waters Run*, 183.

32. Ibid., 183.

33. Ibid., 196.

34. Report of M. L. Mott quoted in Debo, *And Still the Waters Run*, 232–3.

Endnotes 125

35. Donald L. Parman, *Indians and the American West in the Twentieth Century* (Bloomington: Indiana University Press, 1994), Chapter 3; Prucha, *The Great Father*, Chapter 29.

36. Debo, *And Still the Waters Run*, 180.

37. Ibid., 180.

38. Parman, *Indians and the American West*, 16.

39. J. Frazier and F. D. Grouatt to C. D. Carter, BIA CCF, Seminole, File 123, RG 75, NA.

40. J. Frazier and F. D. Grouatt to C. D. Carter, BIA CCF, Seminole, File 123, RG 75, NA.

41. H. L. La Croix, "Work of the District Agents of the Five Tribes," *The Indian School Journal*, April 12, 201–2.

42. La Croix, "Work of the District Agents of the Five Tribes," 201.

43. La Croix, "Work of the District Agents of the Five Tribes," 201.

44. La Croix, "Work of the District Agents of the Five Tribes," 201–2.

45. Resolution of Vinita Retailers Association, April 17, 1912, BIA CCF, Records of the Field Offices, Records Relating to Abolishing district Agents, 1912, Box 1, RG 75, FARC Southwest.

46. J. W. Falkner to BIA Commissioner, February 5, 1912, BIA CCF, Records of the Field Offices, Records Relating to Abolishing district Agents, 1912, Box 1, RG 75, FARC Southwest.

47. J. W. Falkner to BIA Commissioner, February 5, 1912, BIA CCF, Records of the Field Offices, Records Relating to Abolishing district Agents, 1912, Box 1, RG 75, FARC Southwest.

48. J. W. Falkner to BIA Commissioner, February 5, 1912, BIA CCF, Records of the Field Offices, Records Relating to Abolishing district Agents, 1912, Box 1, RG 75, FARC Southwest.

49. R. M. Connell to Sen. Robert L. Owen, Aug. 15, 1912, BIA CCF, Records of the Field Offices, Records Relating to Abolishing district Agents, 1912, Box 1, RG 75, FARC Southwest.

50. Unknown writer to Dana H. Kelsey, United States Indian Superintendent, Aug. 24, 1912, BIA CCF, Records of the Field Offices, Records Relating to Abolishing district Agents, 1912, Box 1, RG 75, FARC Southwest.

51. R. C. Wolfe to Sec. of Interior, Aug. 12, 1912, BIA CCF, Records of the Field Offices, Records Relating to Abolishing district Agents, 1912, Box 1, RG 75, FARC Southwest.

52. R. C. Wolfe to Sec. of Interior, Aug. 11, 1912, BIA CCF, Records of the Field Offices, Records Relating to Abolishing district Agents, 1912, Box 1, RG 75, FARC Southwest.

53. Typescript of "Statement of W. W. Hastings," *The Ardmore Statesman*, April 13, 1912, vol. 6, no. 14, found in Litton, *Cherokee Papers*, 1901–1925, Oklahoma Historical Society, Oklahoma City (OHS), 380.

54. "Statement of W. W. Hastings," 381–2.

55. "Statement of W. W. Hastings," 380.

56. "Statement of W. W. Hastings," 383.

57. M. L. Mott to Father Ketchum, Aug. 16, 1912, BIA CCF, Records of the Field Offices, Records Relating to Abolishing district Agents, 1912, Box 1, RG 75, FARC Southwest.

58. M. L. Mott to Father Ketchum, Aug. 16, 1912, BIA CCF, Records of the Field Offices, Records Relating to Abolishing district Agents, 1912, Box 1, RG 75, FARC Southwest.

59. W. C. Rogers to Dana H. Kelsey, Dec. 13, 1911, BIA CCF, Records of the Field Offices, Records Relating to Abolishing district Agents, 1912, Box 1, RG 75, FARC Southwest.

60. Parman, *American Indians and the West*, 16–17.

61. Prucha, *The Great Father*, 879–84; Parman, *American Indians and the West*, 16–17.

62. Prucha, *The Great Father*, 770.

63. Ibid., 880.

64. McDonnell, *The Dispossession of the American Indian*; McDonnell, "Competency Commissions and Indian Land Policy, 1913–1920," *South Dakota History* 11 (1980).

65. Gabe E. Parker to Commissioner of Indian Affairs, June 1, 1918, BIA CCF, Five Tribes Agency, Records of Field Offices, Records of Supervising Field Clerks, Box 2, Competency Commission Folder, RG 75, FARC Southwest.

66. Gabe E. Parker to Commissioner of Indian Affairs, June 1, 1918, BIA CCF, Five Tribes Agency, Records of Field Offices, Records of Supervising Field Clerks, Box 2, Competency Commission Folder, RG 75, FARC Southwest.

67. O. M. McPherson to Commissioner of Indian Affairs, April 12, 1917, BIA CCF, Five Tribes Agency, Records of Field Offices, Records of Supervising Field Clerks, Box 2, Competency Commission Folder, RG 75, FARC Southwest.

68. J. Marcus to Gabe E. Parker, October 18, 1917, BIA CCF, Five Tribes Agency, Records of Field Offices, Records of Supervising Field Clerks, Box 2, Competency Commission folder, RG 75, FARC Southwest.

69. Gabe E. Parker to Commissioner of Indian Affairs, June 1, 1918, BIA CCF, Five Tribes Agency, Records of Field Offices, Records of Supervising Field Clerks, Box 2, Competency Commission Folder, RG 75, FARC Southwest

70. Gabe E. Parker to Commissioner of Indian Affairs, June 1, 1918, BIA CCF, Five Tribes Agency, Records of Field Offices, Records of Supervising Field Clerks, Box 2, Competency Commission Folder, RG 75, FARC Southwest.

71. Gabe E. Parker to Commissioner of Indian Affairs, June 1, 1918, BIA CCF, Five Tribes Agency, Records of Field Offices, Records of Supervising Field Clerks, Box 2, Competency Commission Folder, RG 75, FARC Southwest.

72. Cox to Robert L. Williams, February 13, 1917, Robert L. Williams Papers, Box 7, OHS. Several months earlier, the governor received a letter from an attorney informing him of a recently alienated allotment: "If you are interested in buying the land...advise me what you are able to pay and I will endeavor to secure the title for you....I desire to give you the preference and will not agree to represent anyone else until I hear from you." Robert M.

Endnotes 127

Rainey to Hon. R. L. Williams, November 17, 1916, Robert L. Williams Papers, Box 6, OHS.

73. For studies of Indians and military service, see John L. Finger, *Cherokee Americans: The Eastern Band of the Cherokee in the Twentieth Century* (Lincoln: University of Nebraska Press, 1991); Michael L. Tate, "From Scout to Doughboy: The National Debate over Integrating American Indians into the Military, 1891–1918," *Western Historical Quarterly* XVII (1986); W. Bruce White, "The American Indian as Soldier, 1890–1919," *Canadian Review of American Studies* VII (1976); John Whiteclay Chambers II, *To Raise an Army: The Draft Comes to Modern America* (New York: Free Press, 1987); Erik M. Zissu, "Conscription, Sovereignty, and Land: American Indian Resistance During World War I," *Pacific Historical Review* 64 (November, 1995).

74. Parker complained at numerous instances of his dwindling manpower and the urgent nature of providing more help to monitor and mitigate the effects of mobilization upon his charges. Gabe E. Parker to Commissioner of Indian Affairs, May 18, 1917, BIA CCF, Five Tribes Agency, Records of Field Offices, Records of Supervising Field Clerks, Box 2, Competency Commission Folder, RG 75, FARC Southwest.

75. The case was that of Loring Hotinlubbi. Parker's actions on behalf of Hotinlubbi went far beyond the call of duty. There was also another aspect to Hotinlubbi's case that was interesting. Hotinlubbi, who was evidently dark complected, was accused of stabbing the French farmer because the farmer had insulted him by alluding to his dark skin. The farmer, who sold spirits out of his home to soldiers, did not allow blacks to drink inside his house, while he let white soldiers do so. The army prosecutor accused Hotinlubbi of becoming enraged because the farmer would not allow him to drink inside with other soldiers. John C. Wilburn to Gabe E. Parker, August, 1919, BIA CCF, RG 75, NA.

76. Grayson began his letter to Cato Sells, Commissioner of Indian Affairs, by referring to "the continually repeated pathetic requests of our full blood Indians — old men and old women and near relatives of young men whom the government has, in its terrible emergency been forced to impress into its military service for work on the battle fields of Europe." He thus immediately distanced himself, and other progressives, from the conservative Creeks while also seeking to obtain protections for those conservatives. In fact, Grayson himself had served in uniform during the Civil War and his son was an officer during World War I. As for the reasoning behind the conservatives' appeals, he wrote: "I am constrained to present their desires in this matter and ask your very serious consideration of them." "The full blood Creek has an innate aversion and horror of seeing his loved ones sent three thousand miles overseas." Grayson also went on to note the severe difficulties the Creeks would face in the military as their English language skills were rudimentary. G. W. Grayson to Cato Sells, February 18, 1918, BIA CCF, Creek, File 125, RG 75, NA.

77. The document record is littered with acts on behalf of conservative tribal members during the war. Parker in particular worked to gain tribal members equality before the law and in the military.

CHAPTER SIX

1. Angie Debo, *And Still the Waters Run* (Princeton: Princeton University Press, 1940), 321.

2. Ibid., 321. One newspaper account noted that "Superintendent Locke said that politics would have no place in the administration of his office and that the welfare of the Indians will have the first consideration....It is said that Mr. Locke...is not worrying and does not intend that political pressure shall influence him....Superintendent Locke said that politics would have no place in the administration of his office and that the welfare of the Indians will have the first consideration." *Muskogee Daily Phoenix*, June 23, 1921.

3. Locke also engaged his party bosses in a feud throughout June 1921 regarding the hiring of a superintendent of schools. He wished to recruit his predecessor, Gabe Parker, to head up the schools, but Parker was a Democrat. A newspaper report maintained that "[t]he appointment [of Parker] upon its face is not a political one but comes of Mr. Locke's expressed desire to name a man whom he felt was best qualified for the place." *Muskogee Daily Phoenix*, June 20, 1921, November 13, 1921. By the end of June the pressure was too great to ignore. "As the protests which have been so strong and so numerous that it is understood Commissioner Burke can not ignore them." *Muskogee Daily Phoenix*, June 30, 1921.

4. Debo, *And Still the Waters Run*, 321.

5. Francis Paul Prucha, *The Great Father, The United States Government and the American Indians*, Vol. II (Lincoln: University of Nebraska Press, 1984), 759–62.

6. For discussion of shifting trends in the thinking about and implementation of federal Indian policy in the late nineteenth and early twentieth centuries, see Frederick E. Hoxie, *A Final Promise: The Campaign to Assimilate the Indians, 1880–1920* (Lincoln: University of Nebraska Press, 1984), Chapters 3, 4.

7. The objectives of the Indian bureau were consequently dimmed. By the 1910s and into the 1920s, the bureau's accumulated experiences and acquired knowledge suggested that as a race, Indians were victims of their "natural" position in the social hierarchy. Like other "inferior" races, they could expect only so much from life and it was best that they should become inured to the harshness of competition sooner rather than later. Even the revolutionary tilt of the anthropologist Franz Boas, who came to regard cultures outside of the framework of a hierarchy, expected Indians to vanish in the near future, a dying breed unable to adapt to an entirely different civilization. Given the "scientific" wisdom of the day, and the evident failure of the

Endnotes

Dawes Act to jolt Indians into modern society, Indians would simply have to be satisfied with limited futures. Hoxie, *A Final Promise*, 139–43.

8. Prucha, *The Great Father*, 881.
9. Debo, *And Still the Waters Run*, 317.
10. Debo, 318.
11. Both progressives and conservatives fretted over public education and voiced concerns that all Indians would encounter discrimination. Gabe E. Parker to Commissioner of Indian Affairs, April 17, 1918, Office Files of Various Superintendent's 1920s. Box 5, RG 75 FARC Southwest.
12. Daniel F. Littlefield, Jr., *Alex Posey: Creek Poet, Journalist, and Humorist* (Lincoln: University of Nebraska Press, 1992).
13. Clyde Ellis, *To Change Them Forever: Indian Education at the Rainy Mountain Boarding School, 1893–1920* (Norman: University of Oklahoma Press, 1996), 131. Robert A. Trennert, Jr., *The Phoenix Indian School: Forced Assimilation in Arizona, 1891–1935* (Norman: University of Oklahoma Press, 1988); Devon A. Mihesuah, *Cultivating the Rosebuds: The Education of Women at the Cherokee Female Seminary, 1851–1909* (Urbana-Champagne: University of Illinois Press, 1993); K. Tsianina Lomawaima, *They Called it Prairie Light: The Story of Chilocco Indian School* (Lincoln: University of Nebraska Press, 1994); Margaret C. Szasz, *Education and the American Indian: The Road to Self-Determination, 1928–1973* (Albuquerque: University of New Mexico Press, 1974).
14. Lomawaima, *They Called it Prairie Light*, 99.
15. Confirmation of progressives' fears was supplied in 1928 by a government report published by the Brookings Institution. Commonly known as the Meriam Report, named for its principle researcher Lewis Meriam of the Institute for Government Research, this thick compilation of evidence from around the country amounted to an indictment of the Indian Office. Meriam lambasted the government's mismanagement of Indian affairs and recorded the serious decline of the conditions of Indian life. Indians were described as more susceptible to debilitating poverty and poor health than any other group in the nation. Opportunities were woefully lacking, including educational programs. *The Problem of Indian Administration* (Washington, DC: Brookings Institute, 1928).
16. Archival documents and various secondary sources indicate that attacks upon Indians could not be contained during the early 1920s.
17. Nigel Sellars, "Wobblies in the Oil Fields: The Suppression of the Industrial Workers of the World in Oklahoma," in Davis D. Joyce, ed., *'An Oklahoma I Had Never Seen Before': Alternative Views of Oklahoma History* (Norman: University of Oklahoma Press, 1994), 129–44; Marvin E. Kroeker, "'In Death You Shall Not Wear It Either,' The Persecution of Mennonite Pacifists in Oklahoma," in Davis, *'An Oklahoma I had never Seen Before,'* 81–2; Douglas Hale, "The People of Oklahoma: Economics and Social Change, in Anne H. and H. Wayne Morgan, eds., *Oklahoma: New Views of the Forty-Sixth State* (Norman: University of Oklahoma Press, 1982), 53–4; Sheila Manes,

"Pioneers and Survivors: Oklahoma's Landless Farmers," in Morgan eds., *Oklahoma*, 117–20; Garin Burbank, *When Farmers Voted Red: The Gospel of Socialism in the Oklahoma Countryside, 1910–1924* (Westport: Greenwood Press, 1978); James R. Green, *Grass-Roots Socialism: Radical Movements in the Southwest, 1895–1943* (Baton Rouge: Lousiana State University Press, 1978).

18. In 1910, there were nine German-language papers in the state; by 1930 there was one. Hale, "The People of Oklahoma," 41–2. For a discussion of the Socialist Party's difficulties during the war see Manes, "Pioneers and Survivors," 119. The forces of loyalty were well organized, and county-level Councils of Defense sprang up throughout the state, branches of an organization that proliferated throughout the country during the war. See, David M. Kennedy, *Over Here: The First World War and American Society* (New York: Oxford University Press, 1980).

19. For discussion of the Green Corn Rebellion, see Erik M. Zissu, "Conscription, Sovereignty, and Land: American Indian Resistance During World War I," *Pacific Historical Review* 64, (November, 1995); Burbank, *When Farmers Voted Red*, 134; Green, *Grass-Roots Socialism*, 360.

20. See John L. Thompson, *Closing of the Frontier: Radical Response in Oklahoma, 1889–1923* (Norman: University of Oklahoma Press, 1986); Manes, "Pioneers and Survivors," 119.

21. Carter Blue Clark, "A History of the Ku Klux Klan in Oklahoma," (unpublished Ph.D. dissertation, University of Oklahoma, 1976).

22. Scott Ellsworth, *Death in a Promised Land: The Tulsa Race Riot of 1921* (Baton Rouge: Louisiana State University Press, 1982). This was hardly a singular event. Tensions between whites and blacks had been long-running. During World War I, these tensions had further escalated, especially in Texas immediately south of Oklahoma.

23. There is a rich selection of writings about Oklahoma's oil boom period. See, for example, Jerome O. Steffen, "Stages of Development in Oklahoma History," in Morgan, eds., *Oklahoma*, 26–7; Arrell M. Gibson, *The History of Oklahoma* (Norman: University of Oklahoma Press, 1984), and Gibson, *Wilderness Bonanza: The Tri-State District of Missouri, Kansas, and Oklahoma* (Norman: University of Oklahoma Press, 1972).

24. Kenneth S. Johnson, "Minerals, Mineral Industries and Reclamation," in John W. Morris, ed., *Geography of Oklahoma* (Oklahoma City: Oklahoma Historical Society, 1977), 95.

25. Steffen, "Stages of Development," 26–7; Hale, "The People of Oklahoma," 58–9.

26. Johnson, "Minerals, Mineral Industries and Reclamation," 98.

27. Hale, "The People of Oklahoma", 58–9.

28. Underground reservoir is the title of a book by sociologist Terry P. Wilson, *The Underground Reservation: Osage Oil* (Lincoln: University of Nebraska Press, 1985). For more on the Osage oil wealth and its disastrous consequences see Dennis McAuliffe, *The Deaths of Sybil Bolton: An American History* (New York: Times Books, 1994).

Endnotes 131

29. This disillusion, it should be noted, embraced more than individual aspiration. As historians have shown, the bounty of Oklahoma was largely skimmed off by outsiders, by large oil corporations elsewhere. Oklahoma, like other western states rich in mineral deposits, was tethered to the rest of the nation much as a colony. Its riches served to boost the fortunes of other regions and other populations. Although "many Oklahomans became instantly wealthy," one writer notes, "this figure was nothing compared to the staggering amounts of money drained from the...state. While the economic fortunes of Oklahoma were benefited by such men as J. Paul Getty, E. W. Marland, Frank Phillips, William S. Skelly, and Harry Sinclair, who actually lived in the state, the major portion of the capital generated by the oil fields was drained from the region." Owing to the high cost of machinery, leases, exploration, and other expenses, the heyday of the small, maverick oilman was brief, and by the mid-1920's it had been replaced by the consolidation of "integrated major oil companies." In 1924, a handful of companies produced fifty percent of the state's oil. Familiar names were behind these companies and often held substantial and controlling interests in them. The Rockefellers, Mellons, and Pews directed some of the largest oil concerns in Oklahoma. Steffen, "Stages of Oklahoma Development," 27.

30. Wilson, *Underground Reservation*; McAuliffe, *The Deaths of Sybil Bolton*.

31. *Tulsa Tribune*, June 3, 1926; McAuliffe, *The Deaths of Sybil Bolton*, 97–8, 147; Wilson, *Underground Reservation*, 144–7.

32. *Tulsa Tribune*, June 3, 1926.

33. *The Independent*, Vol. 116, Feb. 20, 1926, 227–8.

34. *The Independent*, Vol. 116, Feb. 20, 1926, 227–8.

35. See Wilson, *Underground Reservation*, 121–47.

36. As is apparent, the methods already in use by grafters in the first years after Oklahoma statehood were applied in the oil gambit. They were often applied, however, with greater ferocity in the pursuit of oil. See Debo, *And Still the Waters Run*; Danney Goble, *Progressive Oklahoma: The making of a New Kind of State* (Norman: University of Oklahoma Press, 1980), 78–85.

37. Supervising Field Clerk to A. G. McMillan, Acting Superintendent of the Five Tribes Agency, September 20, 1929 (two reports of the same date), BIA CCF, Records of Supervising Field Clerks, Box 2, Poisoned Choctaws Folder, RG 75, FARC Southwest.

38. Supervising Field Clerk to A. G. McMillan, Acting Superintendent of the Five Tribes Agency, September 20, 1929 (two reports of the same date), BIA CCF, Records of Supervising Field Clerks, Box 2, Poisoned Choctaws Folder, RG 75, FARC Southwest.

39. Supervising Field Clerk to A. G. McMillan, Acting Superintendent of the Five Tribes Agency, September 20, 1929 (two reports of the same date), BIA CCF, Records of Supervising Field Clerks, Box 2, Poisoned Choctaws Folder, RG 75, FARC Southwest.

40. E. B. Merritt, Assistant Indian Bureau Commissioner, to Chickasaw Gov. Douglas Johnston, May 9, 1919, BIA CCF, Chickasaw, File 175, RG 75, NA.

41. Copy in BIA CCF, Records of the Office of the Superintendent, Inspection Case Files, Box 3, Miscellaneous, FARC Southwest, p. 11.

42. Matthew Sniffen, Gertrude Bonnin, Charles H. Fabens, "Oklahoma's Poor Rich Indians: An Orgy of Graft and Exploitation of the Five Civilized Tribes — Legalized Robbery" (Philadelphia: Indian Rights Asscociation, 1924). The author thanks Daniel F. Littlefield, Jr., for providing a Xeroxed copy of the pamphlet.

43. "Oklahoma's Poor Rich Indians," 5.

44. "Oklahoma's Poor Rich Indians," 7.

45. "Oklahoma's Poor Rich Indians," 6.

46. "Oklahoma's Poor Rich Indians," 23–6.

47. "Oklahoma's Poor Rich Indians," 26–7.

48. "Oklahoma's Poor Rich Indians," 27–8.

49. "Oklahoma's Poor Rich Indians," 28.

50. "Oklahoma's Poor Rich Indians," 6.

51. Debo, *And Still the Waters Run*, 327–8.

52. The Jackson Barnett case has been written about by numerous authors, including Debo, Goble and others. Documents relating to the case are voluminous and can be found in the Western History Collection at the University of Oklahoma and FARC Southwest, among other places. Tribal members did have their share of protectors, who worked to shield assets from non-Indians. One attorney developed a complex system of trusts in an effort to help tribal members retain larger portions of their wealth. See Ben Cates Collection, Western History Collection (WHC), University of Oklahoma.

53. See Debo, *And Still the Water Runs*.

54. The notion that American society became one in which organized groups were increasingly necessary in order to advance the interests of individuals and groups has been explored by a number of scholars. Robert Wiebe has asserted that the large change in American society from 1890 to 1920 was one from "island communities" to one marked by "the regulative, hierarchical needs of urban-industrial life." This process not only targeted the relatively powerless, such as Indians. Indeed, big business tended to work at cross purposes with any number of entities, even those as large as states. A number of studies have shown that the largest oil companies, including Standard Oil, gained dominance over the industry and siphoned profits away from the areas in which it was drilled. Oklahoma was treated as a virtual colony by these industrial behemoths, a trend that some writers see as endemic to the entire western region. Robert H. Wiebe, *The Search for Order, 1877–1920* (New York: Hill and Wang, 1967), xiii-xiv; John Whiteclay Chambers, II, *The Tyranny of Change: America in the Progressive Era, 1890–1920* (New York: St. Martin's Press, 1992).

55. *Okmulgee Daily Times*, October 2, 1925, 1.

56. Some historians have derogatorily referred to men and women who participated in groups such as the Apela Club as "Red Progressives," as shallow imitators of the white middle class. This criticism simplistically ignores

Endnotes

the fact that the basis of this organization was a re-vitalized sense of Indianness that was aroused in response to the developments of the early twentieth century. Moreover, such groups reversed the trend of separating the various tribal elements and signaled a shift toward unity and solidarity. Although their methods smacked of Progressive Era gentility, it is necessary to look beyond the form to the substance of their activities. Brian Dippie is among those who dismisses "Red Progressives." See his otherwise insightful study, *The Vanishing American: White Attitudes and U. S. Indian Policy* (Middletown: Wesleyan University Press, 1982).

57. Newspaper clippings, n.d., Lee F. Harkins Papers, Box 3, Scrapbook, Oklahoma Historical Society (OHS), Oklahoma City.

58. Newspaper clippings, n.d., Lee F. Harkins Papers, Box 3, Scrapbook, OHS.

59. Debo, *And Still the Waters Run*, 127.

60. Debo, 330–1.

CHAPTER SEVEN

1. Rachel Caroline Eaton, "The Legend of the Battle of Claremore Mound." Her essay first appeared in the *Chronicles of Oklahoma* in 1930, and is reprinted in Daniel F. Littlefield, Jr., and James W. Parins, eds., *Native American Writing in the Southeast: An Anthology, 1875–1935* (Jackson: University of Mississippi Press, 1995), 238–41.

2. Eaton also suggested an evolutionary model of human development, an idea more relevant to later sections of this chapter. In her retelling of the battle, she attributed Cherokee superiority to their "progressive" nature as compared to the more primitive state of Osage civilization. Biographical information on Eaton is drawn from Littlefield and Parins, eds., *Native American Writing in the Southeast*, 235. Prof. Littlefield has also kindly allowed the author access to his many files on Indian writers kept at the Native American Press Archives at the University of Arkansas at Little Rock.

3. "General Convention of Oklahoma Indians," February 26, 1924, BIA CCF, Office Files of Various Superintendents, 1920s, RG 75, FARC Southwest.

4. An example is that of Ruth Muskrat Bronson, a Cherokee women of mixed descent who dedicated her professional career, first with the Indian bureau and later with the National Congress of American Indians, to Indians. Gretchen G. Harvey, "Cherokee and American: Ruth Muskrat Bronson, 1897–1982," (unpublished Ph.D. dissertation, Arizona State University, 1996).

5. George Todd Downing, "A Choctaw's Autobiography," *American Indian Magazine*, December 1926, 10.

6. Today, paradoxically, blood quantum is of major importance. As the stigma of Indian ancestry has receded in many areas, tribes have been confronted with a new dilemma: more people are claiming to be Indian. As a result, blood quantum has come to occupy a central place in the determina-

tion of membership. As the politics of authenticity have evolved, blood continues to be the main link to tribal membership. Whereas the threats to Indian groups earlier in the century prompted some tribes to be more inclusive, the recent surge in Indian identification has forced them to close ranks and again to rely on a more exclusive definition of Indianness. For more on this vast subject, see, among others, Matthew C. Snipp, *American Indians: The First of this Land* (New York: Russell Sage Foundation, 1989); Joane Nagel, *American Indian Ethnic Renewal: Red Power and the Resurgence of Identity and Culture* (New York: Oxford University Press, 1996); Terry P. Wilson, "Blood Quantum: Native American Mixed Bloods," in Marla P. P. Root, ed., *Racially Mixed People in America* (Newbury Park, CA: Sage Publications, 1992), 108–125; Russell Thornton, *The Cherokees: A Demographic History* (Lincoln: University of Nebraska Press, 1990).

7. Redbird Smith is credited by some scholars with spearheading what they term "the Redbird Smith movement," which depended to a significant degree upon Smith's power and reputation as a leader. See Robert K. Smith, The Redbird Smith Movement" (M.A. Thesis, University of Arizona, 1953). A microfilm copy of the thesis exists in the Western History Collection (WHC), University of Oklahoma.

8. Levi Gritts, "Legend of Keetoowah Society, Nov. 14, 1930, document in Indian Pioneer History, Cherokee, Keetoowah Society, Oklahoma Historical Society (OHS), photocopy obtained from Daniel Littlefield. Jr. [The Indian Pioneer History consulted for this study included both the copy at the OHS and the one in the WHC, which is called Indian Pioneer Papers. Since these copies have different indexes and separate volume orders and pagination, I have specified which collection is cited.]

9. "Illinois Fire, November 10, 1920," Lewis, S. R. Documents, Indian Pioneer History, Vol 6, 240, OHS.

10. "Illinois Fire, November 10, 1920," Lewis, S. R. Documents, Indian Pioneer History, Vol 6, 240, OHS.

11. "Illinois Fire, November 10, 1920," Lewis, S. R. Documents, Indian Pioneer History, Vol 6, 240–1, OHS.

12. "Illinois Fire, November 10, 1920," Lewis, S. R. Documents, Indian Pioneer History, Vol 6, 240–1, OHS.

13. "Illinois Fire, November 10, 1920," Lewis, S. R. Documents, Indian Pioneer History, Vol 6, 241, OHS.

14. "Illinois Fire, November 10, 1920," Lewis, S. R. Documents, Indian Pioneer History, Vol 6, 241, OHS.

15. "Illinois Fire, November 10, 1920," Lewis, S. R. Documents, Indian Pioneer History, Vol 6, 241, OHS.

16. Quoted in *The Tahlequah Arrow-Democrat*, February 4, 1921, 1.

17. *The Tahlequah Arrow-Democrat*, February 4, 1921, 1.

18. Thomas, "The Redbird Smith Movement," 201.

19. *The Tahlequah Arrow-Democrat*, February 4, 1921, 1.

20. *The Tahlequah Arrow-Democrat*, February 4, 1921, 1.

Endnotes 135

21. *Daily Oklahoman*, October 23, 1921, B2:2-3.

22. Levi Gritts to Victor Locke, Jr., June 17, 1921, Records of the Office of the Superintendent, General Correspondence 1921, 1930, Box 1, Five Tribes Agency, RG 75, FARC Southwest.

23. "Illinois Fire, November 10, 1920," Lewis, S. R. Documents, Indian Pioneer History, Vol 6,241.

24. The Chickasaws and Choctaws, owing to their close proximity in the Southeast, had been forced into close relations with one another by the United States during the Removal crisis of the early nineteenth century. This closeness persisted in Indian Territory, and the two tribes were granted joint ownership over large expanses of territory, including profitable and much coveted coal fields and timber lands.

25. This issue touched a sensitive nerve among conservatives who feared a further dilution of the tribe by non-Indians who claimed membership. Time and again the Chickasaws, as well as other tribes, resisted efforts to re-open the Dawes Rolls. Although legitimate claims were made, the threat of opening the rolls was deemed greater than the injustice of excluding some people with strong cases for citizenship. Among the other tribes, the occasional suggestion of re-opening the rolls was similarly deemed a dangerous threat.

26. Johnston to Chickasaws convened at Twinpond, July 14, 1924, Douglas Johnston Papers, OHS [hereafter referred to as Johnston Papers].

27. Johnston to Chickasaws convened at Twinpond, July 14, 1924, Johnston Papers.

28. Johnston to Cravatt, April 3, 1927, Johnston Papers.

29. Johnston to Leavitt, April 5, 1928, Johnston Papers.

30. Cornish to Johnston, March 28, 1928, Johnston Papers.

31. Cornish to Johnston, March 28, 1928, Johnston Papers.

32. Later, the Chickasaws joined forces with their neighbors in creating the Choctaw-Chickasaw Protective Association.

33. Resolutions passed by the Chickasaw Tribe in Convention on November 11, 1929, Johnston Papers.

34. Resolutions passed by the Chickasaw Tribe in Convention on November 11, 1929, Johnston Papers.

35. One New Deal program that found supporters among the Seminoles and other tribes was a loan fund that allowed tribal members to form corporations, often based upon an older tribal town configuration. Among the Creeks, for example, groups of tribal members proved effective organizers and obtained significant monies with which to improve or buy lands. W. David Baird, "Are there Real Indians in Oklahoma?: Historical Perceptions of the Five Civilized Tribes," *Chronicles of Oklahoma* 68, (Spring, 1990). Graham D. Taylor, *The New Deal and American Indian Tribalism: The Administration of the Indian Reorganization Act, 1934-45* (Lincoln: University of Nebraska Press, 1980); Kenneth R. Philp, *John Collier's Crusade for Indian Reform, 1920-1954* (Tucson: University of Arizona Press, 1977).

36. Charles Wisdom, "Report on the Social Condition of the Oklahoma Seminole," 1937, BIA CCF, Records of the Five Civilized Tribes, FARC, Southwest, Seminole Box 60, Anthropological Study of Seminoles File, 14.

37. See Rebecca Belle Bateman for discussion of Seminole-Seminole Freedmen animosity and political separation, "'Were Still Here': History, Kinship, and Group Identity Among the Seminole Freedmen of Oklahoma," (unpublished Ph.D. dissertation, Johns Hopkins University, 1991).

38. Wisdom, "Report on the Social Condition of the Oklahoma Seminole," 13–4.

39. Convention pamphlet, FARC, and Lee F. Harkins Collection, OHS; see also Daniel Littlefield, Jr., "Society of Oklahoma Indians," in Armand S. La Potin, ed., *Native American Voluntary Organizations*, Westport, Conn., 1987. Although scholars have largely minimized the role of progressive organizations, dismissing them as created by dilettantes who dabbled in Indian affairs almost as a hobby and who were isolated from the lives of most tribal members, conservatives often responded positively to these initiatives. By the 1920s, they were increasingly willing to listen to the ideas of men and women who worked to alleviate tribal problems, and they admired the insistent rhetoric that flowed from the various progressive organizations. For criticism of "Red Progressives," see Brian Dippie, *The Vanishing American: White Attitudes and U. S. Indian Policy* (Middletown: Wesleyan University Press, 1982), 263–9.

40. "General Convention of Oklahoma Indians," February 26, 1924, BIA CCF, Office Files of Various Superintendents, 1920s, RG 75, FARC Southwest, 8

41. "General Convention of Oklahoma Indians," February 26, 1924, BIA CCF, Office Files of Various Superintendents, 1920s, RG 75, FARC Southwest, 13.

42. "General Convention of Oklahoma Indians," February 26, 1924, BIA CCF, Office Files of Various Superintendents, 1920s, RG 75, FARC Southwest, 14.

43. "General Convention of Oklahoma Indians," February 26, 1924, BIA CCF, Office Files of Various Superintendents, 1920s, RG 75, FARC Southwest, 14.

44. Together with other Society functions, the encampments provided jarring but pleasant juxtapositions. For example, at the 1927 convention held at Pawhuska, in Osage County, an Indian golf tournament preceded the meeting. *The American Indian*, May, 1927.

45. The appearance of conservatives at the conventions also seemed to fulfill the needs of progressive tribal members, who were pleased to receive the public support of their tribal fellows. Additionally, the presence of fullblood encampments in the midst of the middle-class atmosphere of the conventions soothed the sensitivity that many progressives possessed in regard to their own tentative, even suspect, claims to Indianness. More discussion on this subject follows below.

Endnotes

46. "General Convention of Oklahoma Indians," February 26, 1924, BIA CCF, Office Files of Various Superintendents, 1920s, RG 75, FARC Southwest, 5–6.

47. "General Convention of Oklahoma Indians," February 26, 1924, BIA CCF, Office Files of Various Superintendents, 1920s, RG 75, FARC Southwest, 13.

48. "General Convention of Oklahoma Indians," February 26, 1924, BIA CCF, Office Files of Various Superintendents, 1920s, RG 75, FARC Southwest, 16.

49. "General Convention of Oklahoma Indians," February 26, 1924, BIA CCF, Office Files of Various Superintendents, 1920s, RG 75, FARC Southwest, 2–3.

50. Resolution dated June 15, 1927 at Pawhuska, OK., Reprinted in *The American Indian*, June 1927.

51. Resolution dated June 15, 1926 at Muskogee, OK., BIA CCF, Office Files of Various Superintendents, 1920s, RG 75, FARC Southwest.

52. "A Resolution Memorializing Congress to Repeal and Amend Certain Provisions of Acts, and to Defeat Certain Purposed Legislation Affecting the Restricted Members of the Five Civilized Tribes," BIA CCF, Office Files of Various Superintendents, 1920s, Box 12, RG 75, FARC Southwest.

53. In a report later compiled by the Board of Indian Commissioners, Chandler was described as a former disgruntled employee who had been dismissed from his position at the Quapaw Agency "for cause." Report is partially re-printed in a Department of the Interior "Memorandum for the Press," January 28, 1926, BIA CCF, Inspection Case Files, Records of the Office of the Superintendent, Box 3, Miscellaneous, RG 75, FARC Southwest.

54. Card found in Records of the Office of the Superintendent, BIA CCF, Office Files of Various Superintendents, 1920s, Box 10, O.K. Chandler folder, RG 75, FARC Southwest.

55. Remarks referred to in report by Probate Clerk-Investigator to Samuel Blair, Investigator with the Department of the Interior, August 1925, BIA CCF, Records of the Office of the Superintendent, Office Files of Various Superintendents 1920s, Box 10, O. K. Chandler Folder, RG 75, FARC Southwest.

56. Resolutions passed by Nighthawk Keetoowah Society at Gore, OK., Oct. 10, 1925, BIA CCF, Records of the Office of the Superintendents, Box 12, RG 75, FARC Southwest.

57. Probate Clerk Investigator to Samuel Blair, Investigator with the Department of the Interior, August 1925, BIA CCF, Records of the Office of the Superintendent, Office Files of Various Superintendents 1920s, Box 10, O. K. Chandler Folder, RG 75, FARC Southwest.

58. *Muskogee Daily Phoenix*, August 16, 1925.

59. Records of the Office of the Superintendent, BIA CCF, Office Files of Various Superintendents, 1920s, Box 10, O.K. Chandler folder, RG 75, FARC Southwest.

60. *Muskogee Daily Phoenix*, August 16, 1925.
61. *Muskogee Daily News*, January 16, 1926.
62. Littlefield and Parins, eds., *Native American Writing in the Southeast*, xviii.
63. Littlefield and Parins, xx, xxi.
64. Mabel Washbourne Anderson, "General Stand Watie," in Littlefield and Parins, eds., *Native American Writing in the Southeast*, (re-printed from *The Chronicles of Oklahoma*), 52–9.
65. Anderson, "General Stand Watie," 55.
66. Anderson, 55.
67. Anderson, 55–9.
68. Francis P. Prucha, *The Great Father: The United States Government and the American Indians*, Vol. II (Lincoln: University of Nebraska Press, 1984), 790. According to another writer, "[t]he 1920s and 1930s marked crucial milestones in United States Indian policy and administration. In the 1920s, liberal interest groups in the East that had been founded in the late 1800s and early 1900s to promote Native American rights and education began in earnest a movement to reform the U.S. Office of Indian Affairs." K. Tsianina Lomawaima, *They Called it Prairie Light: The Story of Chilocco Indian School* (Lincoln: University of Nebraska Press, 1994), 6–7.
69. Prucha, *The Great Father*, 790.
70. Criticism of Indian policy and administration culminated in 1928 with the publication of the Meriam Report. This massive report was nothing short of an indictment of the Indian bureau's mismanagement. Lewis Meriam, *The Problem of Indian Administration* (Washington, DC: The Brookings Institution, 1928).
71. Prucha, *The Great Father*, 798–800.
72. Prucha, 908–9.
73. The literature on artistic and intellectual attitudes toward native peoples during the early twentieth century is growing. See, Carter Jones, "'Hope for the Race of Man': Indians, Intellectuals and the Regeneration of Modern America, 1917–1934," (unpublished Ph.D. dissertation, Brown University, 1991). Also, Walter Benn Michaels, *Our America: Nativism, Modernism, and Pluralism* (Durham: Duke University Press, 1994).
74. Jones, "'Hope for the Race of Man,'" 62–92, 140–53. Also, Philp, *John Collier's Crusade for Indian Reform*. E. A. Schwartz offers a critical view of Collier for patronizing Indian groups. "Red Atlantis Revisited: Culture and Community in the Writings of John Collier," *American Indian Quarterly* 18 (Fall 1994): 507–31.
75. Sunshine Rider, "The Harmony of Nature — 'Why Evergreens are Green,'" in *The American Indian*, January 1927, 7.
76. *The American Indian*, vol. 1, no. 1, 1926.
77. Letter to *The American Indian*, 1927.
78. Letter to *The American Indian*, 1927.

79. John M. Oskison, "The Passing of the Old Indian," *Munsey's Magazine*, 51, April 1914, 527.
80. Oskison, "The Passing of the Old Indian," 528.
81. Oskison, "The Passing of the Old Indian," 529–30, 532–3.
82. Oskison, "The Passing of the Old Indian," 529–30.
83. James A. Davis, "Sequoyah is Called 'Intellectual Genius' of New World, *The American Indian*, January 1927, 14–5. Speculation as to why Sequoyah was thought to be of European ancestry derived from Sequoyah's being given the name George Guess in some circles.
84. See Jones, "'Hope for the Race of Man.'"
85. Oskison, "The Passing of the Old Indian," 535.
86. Downing, "A Choctaw's Autobiography," 10.
87. Downing, "A Choctaw's Autobiography," 10. Mindful of how the region had changed in the decades since his grandfather arrived from Mississippi, Downing sketched his autobiography within the broader frame of Choctaw history, so as "to make it both more interesting and instructive." By interweaving family and tribal history, Downing neatly underlined the radical differences between grandfather and grandson and the ways they exercised Indian identity. Downing understood why the path taken by his grandfather to move in among the Choctaws and come to represent them, to become, in other words, Choctaw, belonged to an Indian past. Behavior made it possible for his grandfather to become Indian. Acting in the best interests of the tribe, taking an Indian wife, and living among tribal members all were favorable characteristics from the viewpoint of fellow Choctaws. For Downing, however, behavior no longer was the arbiter of Indianness. Instead, blood was his strongest link to an Indian identity.
88. Oskison, "The Passing of the Old Indian," 523.
89. Oskison, "The Passing of the Old Indian," 523–4.
90. "Indians' Culinary Art Greatly Aided American Colonists," *The American Indian*, July 1927, 13.
91. "Daughter of Chief Expresses 'Sentiments of a Seminole," *The American Indian*, October 1926, 5.
92. Charles D. Carter, "Memorable Debate Between Pushmataha and Tecumseh," re-printed in *The American Indian*, October 1926, 14–5.
93. Carter, "Memorable Debate between Pushmataha and Tecumseh," 14.

CHAPTER EIGHT

1. For Collier and his New Deal for the Indians see Kenneth R. Philp, *John Collier's Crusade for Indian Reform, 1920–1954* (Tucson: University of Arizona Press, 1977); Graham D. Taylor, *The New Deal and American Indian Tribalism: The Administration of the Indian Reorganization Act, 1934–45* (Lincoln: University of Nebraska Press, 1980).
2. In the accounts of several of the meetings, tribal representatives are documented as both cautiously criticizing Collier and offering outright oppo-

sition to his plans. Testimony given before John Collier and Sen. Elmer Thomas, Collection of The Hon. Elmer Thomas, Subject A, Box 9, Folder A 77, Carl Albert Congressional Research and Studies Center Congressional Archives, University of Oklahoma (Elmer Thomas Papers). Also W. David Baird, "Are There 'Real' Indians in Oklahoma?: Historical Perceptions of the Five Civilized Tribes," *Chronicles of Oklahoma*, 68, no. 1 (1990): 4–18.

3. Creeks later made use of funds to form "corporations." Additionally, Collier received support, albeit guarded and with contingencies attached, from some tribal representatives. Testimony given before John Collier and Sen. Elmer Thomas, Subject A, Box 9, Folder A 77, Elmer Thomas Papers.

4. For the Oklahoma Indian Welfare Act, see Philp, *John Collier's Crusade for Indian Reform*, and Taylor, *The New Deal and American Indian Tribalism*.

5. A constitutional crisis has flared during the past year among the Cherokees.

6. E. A. Scwhartz, "Red Atlantis Revisited: Community and Culture in the Writings of John Collier," *American Indian Quarterly* 18 (Fall, 1994): 515–16, 520–25.

Bibliography

MANUSCRIPT COLLECTIONS AND ARCHIVAL MATERIALS
National Archives, Washington, D.C.

Record Group 75
   Bureau of Indian Affairs Central Classified Files, 1907–1957
      Records of:
         Cherokee Nation
         Chickasaw
         Chilocco
         Choctaw
         Creek
         Five Tribes Agency
         Indian Office
         Seminole

   Division of Information Records Relating to Indians in World War I and World War II, 1920–1921; 1945

Federal Archives and Record Center, Southwest Region, Fort Worth, Texas.

Record Group 75
   Bureau of Indian Affairs Central Classified Files, 1907–1957
      Records of the Five Civilized Tribes Agency
         General Correspondence, 1921, 1930
         Letters Received, 1909–1925

Office Files of Various Superintendents, 1920s
Office Files of Tribal Affairs Officer, Choctaw
Office of Tribal Operations
Records of the Field Offices
Records of the Office of the Superintendent
Records Relating to Veterans
Records of Supervising Field Clerks
Seminole

National Archives, College Park, Maryland

Record Group 48
   Office of the Secretary of the Interior

National Archives, Suitland, Maryland

Record Group 163
   Selective Service System, 1917–1919
      Local Board Experience Files
      States File, Oklahoma

Native American Press Archives, University of Arkansas, Little Rock, Little Rock, Arkansas.

Assorted papers and files

Oklahoma Historical Society, Oklahoma City, Oklahoma.

Indian Pioneer History
Lee F. Harkins Collection
Douglas Johnston Papers
Robert L. Williams Collection
Muriel H. Wright Collection

Western History Collection, University of Oklahoma, Norman, Oklahoma.

John Alley Collection
Frank J. Boudinot Collection
C. S. Cate Collection
Cherokee Nation Papers
G. N. Belvin Collection
Grayson Family Papers
Indian Pioneer Papers
Mrs. Alfred Mitchell Collection
Morris Edward Opler Collection
John M. Oskinson Collection
Moty Tiger Collection

Bibliography

Carl Albert Congressional Research and Studies Center

Congressional Archives
Thomas A. Chandler Collection
Elmer Thomas Collection
L. M. Gensman Collection

University of Tulsa

McFarlin Library
    Typescript, William G. McLoughlin, "The Cherokee Keetoowah Society and the Coming of the Civil War."

GOVERNMENT DOCUMENTS

H. Doc. 409 Annual Report of the Commissioner of Indian Affairs, 1919
H. Doc. 849 Annual Report of the Board of Indian Commissioners, June 30, 1920
H. R. 1133 Indians of the United States, Field Investigation Dec. 18, 1920

NEWSPAPERS AND CONTEMPORARY PERIODICALS

The American Indian
Cherokee Advocate
Daily Chieftain
Daily Oklahoman
The Indian School Journal
Muskogee Phoenix
Muskogee Times Democrat
Okmulgee Daily Times
Tahlequah Arrow
Tulsa Daily World

PAMPHLETS

Bonnin, Gertrude, Fabens, Charles H., Sniffen, Matthew K. *Oklahoma's Poor Rich Indians: An Orgy of Graft and Exploitation of the Five Civilized Tribes—Legalized Robbery.* Philadelphia, Pa.: Office of the Indian Rights Association, 1924.

SECONDARY SOURCES
Unpublished Materials

Bateman, Rebecca Belle. "'We're Still Here': History, Kinship, and Group Identity Among the Seminole Freedmen of Oklahoma." Ph.D. dissertation, The Johns Hopkins University, 1991.
Clark, Carter Blue. "A History of the Ku Klux Klan in Oklahoma." Ph.D. dissertation, University of Oklahoma, 1977.

Harvey, Gretchen Grace. "Cherokee and American: Ruth Muskrat Bronson, 1897–1982." Ph.D. dissertation, Arizona State University, 1996.
Jones, Carter. "'Hope for the Race of Man': Indians, Intellectuals and the Regeneration of Modern American, 1917–1934." Ph.D. dissertation, Brown University, 1991.
May, Katja Helma. "Collision and Collusion: Native Americans and African Americans in the Cherokee and Creek Nations, 1830s to 1920s." Ph.D. dissertation, University of California at Berkeley, 1994.
McIntosh, Kenneth Waldo. "Chitto Harjo, the Crazy Snakes and the Birth of Indian Political Activism in the Twentieth Century." Ph.D. dissertation, Texas Christian University, 1993.
Scales, James R. "Political History of Oklahoma, 1907– 1949." Ph.D. dissertation, University of Oklahoma, 1949.
Thomas, Robert K. "The Redbird Smith Movement." MA thesis, University of Arizona, 1953.
Wickett, Murray R. "Contested Territory: Whites, Native Americans, and African-Americans in Oklahoma, 1865– 1907." Ph.D. dissertation, University of Toronto, 1996.

Articles and Book Chapters

Baird, W. David. "Are there Real Indians in Oklahoma?: Historical Perceptions of the Five Civilized Tribes." *Chronicles of Oklahoma*, 68 (Spring, 1990):4–23.
Bolster, Mel H. "The Smoked Meat Rebellion." *Chronicles of Oklahoma* 31 (1953): 37–55.
Champagne, Duane. "Change, Continuity, and Variation in Native American Societies as a Response to Conquest." In Taylor, William B., and Franklin Pease, eds. *Violence, Resistance, and Survival in the Americas: Native Americans and the Legacy of Conquest.* Washington, DC: Smithsonian Institution Press, 1994: 208–25.
Hagan, William T. "Full Blood, Mixed Blood, Generic, and Ersatz: The Problem of Indian Identity". *Arizona and the West* 27 (Winter, 1985): 309–326.
Herring, Sydney. "Crazy Snake and the Creek Struggle for Sovereignty: The Emergence of a Creek Legal Culture." *American Journal of Legal History* 365 (October, 1990).
Horsman, Reginald. "Scientific Racism and the American Indian in the Mid-Nineteenth Century." *American Quarterly* 27 (May, 1975): 152–168.
Littlefield, Daniel F., Jr. "Utopian dreams of the Cherokee Fullbloods, 1890-1930." *Journal of the West*, 10 (Spring, 1971): 404–427.
———, and Lonnie E. Underhill. "The'Crazy Snake Uprising' of 1909: A Red, Black, or White Affair?" *Arizona and the West* 20 (Winter 1978): 307–324.
Grinde, Donald A., Jr., and Quintard Taylor. "Red vs. Black: Conflict and

Accommodation in Post-Civil War Indian Territory, 1865–1907." *American Indian Quarterly* 8 (Summer, 1984): 211–229.
McDonnell, Janet. "Competency Commissions and Indian Land Policy, 1913–1920." *South Dakota History* 11 (1980): 21–34.
Miles, George. "To Hear and Old Voice: Rediscovering Native Americans in American History." In Cronon, William, George Miles, and Jay Gitlin, eds., *Under an Open Sky: Rethinking America's Western Past*. New York: W. W. Norton, 1992: 52–70.
Oskinson, John M. "The Passing of the Old Indian." *Munsey's Magazine*, 51 (April, 1914): 519–35.
Parker, Gabe E. "The Indian as the Social Equal of the White Man." In Beaulieu, Irene C. and Woodward, Kathleen, *Tributes to a Vanishing Race*. Chicago: privately printed, 1916: 40–2.
Schwartz, E. A. "Red Atlantis Revisited: Community and Culture in the Writings of John Collier." *American Indian Quarterly*. 18 (Fall, 1994): 507–531.
Wilson, Terry P. "Blood Quantum: Native American Mixed Bloods," in Marla P. P. Root, ed., *Racially Mixed People in America*, Newbury Park, CA: Sage Publications, 1992: 108–125.
Wood, David. "American Indian Farmland and the Great War," *Agricultural History* 55 (July, 1981): 249–65.
Wright, Peter M. "John Collier and the Oklahoma Indian Welfare Act of 1936." *Chronicles of Oklahoma* 50 (Autumn, 1972): 347–71.

Books

Bailey, M. Thomas. *Reconstruction in Indian Territory: A Story of Avarice, Discrimination, and Opportunism*. Port Washington, NY: Kennikat Press, 1972.
Baird, W. David, ed. *A Creek Warrior for the Confederacy: The Autobiography of Chief G. W. Grayson*. Norman: University of Oklahoma Press, 1988.
Berkhofer, Richard F. Jr. *The White Man's Indian: Images of the American Indian from Columbus to the Present*. New York: Vintage, 1979.
Burbank, Garin. *When Farmers Voted Red: The Gospel of Socialism in the Oklahoma Countryside, 1910–1924*. Westport: Greenwood Press, 1976.
Burton, Jeffrey. *Indian Territory and the United States, 1866–1906: Courts, Government, and the Movement for Oklahoma Statehood*. Norman: University of Oklahoma Press, 1995.
Champagne, Duane. *Social Order and Political Change: Constitutional Governments Among the Cherokee, the Choctaw, the Chickasaw, and the Creek*. Stanford: Stanford University Press, 1992.
Cornell, Stephen. *The Return of the Native: American Indian Political Resurgence*. New York: Oxford University Press, 1988.
Crockett, Norman L. *The Black Towns*. Lawrence: University of Kansas Press, 1979.

Debo, Angie. *And Still the Waters Run*. Princeton, NJ: Princeton University Press, 1940.

———. *The Road to Disappearance: A History of the Creek Indians*. Norman: University of Oklahoma Press, 1941.

Dippie, Brian W. *The Vanishing American: White Attitudes and U.S. Indian Policy*. Middletown, CT: Wesleyan University Press, 1982.

Ellsworth, Scott. *Death in a Promised Land: The Tulsa Race Riot of 1921*. Baton Rouge: Louisiana State University Press, 1982.

Foreman, Grant. *Indian Removal: The Emigration of the Five Civilized Tribes of Indians*. Norman: University of Oklahoma Press, 1972.

Franklin, Jimmie L. *Journey Toward Hope: A history of Blacks in Oklahoma*. Norman: University of Oklahoma Press, 1982.

Gibson, Arrell M. *The Chickasaws*. Norman: University of Oklahoma Press, 1972.

———. *The History of Oklahoma*. Norman: University of Oklahoma Press, 1984.

Goble, Danney. *Progressive Oklahoma: The Making of a New Kind of State*. Norman: University of Oakhoma Press, 1980.

Green, James R. *Grass-Roots Socialism: Radical Movements in the Southwest, 1895–1943*. Baton Rouge: Louisiana State University Press, 1978.

Hertzberg, Hazel W. *The Search for an American Indian Identity: Modern Pan-Indian Movements*. Syracuse: Syracuse University Press, 1971.

Hoxie, Frederick E. *A Final Promise: The Campaign to Assimilate the Indians, 1880–1920*. Lincoln: University of Nebraska Press, 1984.

Joyce, Davis D., ed. *"An Oklahoma I Had Never Seen Before": Alternative Views of Oklahoma History*. Norman: University of Oklahoma Press, 1994.

Kidwell, Clara Sue. *Choctaws and Missionaries in Mississippi, 1818–1918*. Norman: University of Oklahoma Press, 1995.

King, Duane H., ed. *The Cherokee Indian Nation: A Troubled History*. Knoxville: University of Tennessee Press, 1979.

La Potin, Armand S., ed. *Native American Voluntary Organizations*. Westport, CT: Greenwood Press, 1987.

Leeds, Georgia Rae. *The United Keetoowah Band of Cherokee Indians in Oklahoma*. New York: Peter Lang, 1996.

Littlefield, Daniel F., Jr. *Alex Posey: Creek Poet, Journalist, and Humorist*. Lincoln: University of Nebraska Press, 1992.

———. *Seminole Burning: A Story of Racial Vengeance*, Jackson: University of Mississippi Press, 1996.

———. *The Chickasaw Freedmen: A People Without a Country*. Westport, CT: Greenwood Press, 1980.

———. *The Cherokee Freedmen: From Emancipation to American Citizenship*. Westport: Greenwood Press, 1978.

———, and James W. Parins, eds. *Native American Writing in the Southeast: An Anthology, 1875–1935*. Jackson: University of Mississippi Press, 1995.

Lomawaima, K. Tsianina. *They Called it Prairie Light: The Story of Chilocco*

Indian School. Lincoln: University of Nebraska Press, 1994

May, Katja. *African Americans and Native Americans in the Creek and Cherokee Nations, 1830s to 1920s: Collision and Collusion.* New York: Garland Publishing, 1996,

McAuliffe, Dennis. *The Deaths of Sybil Bolton: An American History.* New York: Times Books, 1994.

McDonnell, Janet. *The Dispossession of the American Indian, 1887–1934.* Bloomington: Indiana University Press, 1991.

McLoughlin, William G. *After the Trail of Tears: The Cherokees' Struggle for Sovereignty, 1839–1880*, Chapel Hill, University of North Carolina Press, 1993.

———. *Cherokee Renascence in the New Republic.* Princeton: Princeton University Press, 1986.

———, and Walter H. Conser, ed. *The Cherokees and Christianity, 1794–1870: Essays on Acculturation and Cultural Persistence.* Athens: University of Georgia Press, 1994.

Meriam, Lewis. *The Problem of Indian Administration.* Washington, DC: Brookings Institution, 1928.

Michaels, Walter Benn. *Our America: Nativism, Modernism, and Pluralism.* Durham: Duke University Press, 1994.

Mihesuah, Devon A. *Cultivating the Rosebuds: The Education of Women at the Cherokee Female Seminary, 1851–1909.* Urbana-Champagne: University of Illinois Press, 1993.

Miner, H. Craig. *The Corporation and the Indian: Tribal Sovereignty and Industrial Civilization in Indian Territory, 1865–1907.* Columbia: University of Missouri Press, 1976.

Moore, John H., ed. The Political Economy of North American Indians. Norman: University of Oklahoma Press, 1993.

Morgan, Anne Hodges, and H. Wayne Morgan, eds. *Oklahoma: New Views of the Forty-Sixth State.* Norman: University of Oklahoma Press, 1982.

Morris, John W, ed., *Geography of Oklahoma.* Oklahoma City: Oklahoma Historical Society, 1977.

Nagel, Joane. *American Indian Ethnic Renewal: Red Power and the Resurgence of Identity and Culture.* New York: Oxford University Press, 1996.

Parman, Donald L. *Indians and the American West in the Twentieth Century.* Bloomington: Indiana University Press, 1994.

Prucha, Francis Paul. *The Great Father: The United States Government and the American Indians.* Vol. II. Lincoln: University of Nebraska Press, 1984.

Rister, Carl C. *Land Hunger: David L. Payne and the Oklahoma Boomers.* New York: Arno Press, 1975.

Stein, Howard F., and Robert F. Hill, eds. *The Culture of Oklahoma.* Norman: University of Oklahoma Press, 1993.

Strickland, Rennard. *The Indians in Oklahoma.* Norman: University of Oklahoma Press, 1980.

Szasz, Margaret C. *Education and the American Indian: The Road to Self-*

Determination, 1928–1973. Albuquerque: University of New Mexico Press, 1974.

Thompson, John L. *The Closing of the Frontier: Radical Response in Oklahoma, 1889–1923*. Norman: University of Oklahoma Press, 1986.

Thornton, Russell. *The Cherokees: A Population History*. Lincoln: University of Nebraska Press, 1990.

Tolson, Arthur L. *The Black Oklahomans: A History, 1541– 1972*. New Orleans: Edwards Printing Co., 1972.

Unrau, William E. *Mixed Bloods and Tribal Dissolution: Charles Curtis and the Quest for Indian Identity*. Lawrence: University of Kansas Press, 1989.

Wardell, Morris L. *A Political History of the Cherokee Nation, 1838–1907*. Norman: University of Oklahoma Press, 1938.

Wilson, Terry P. *The Underground Reservation: Osage Oil*. Lincoln: University of Nebraska Press, 1985.

Index

African Americans
    Oklahoma, 44–45
    poverty, 51–54, 59, 66–67
    relations with Indians, 6, 76, 84, 94–96
    violence against, 45, 76
allotment, 20–21, 24, 32, 38–39 41, 64, 73
American Indian, 102, 104–105
American Indian Defense Association, 81, 101
Anderson, Mabel W., 100
Apela Club, 84–85, 95

blacks, *see* African Americans
Bonnin, Gertrude, 81–83
Boudinot, Frank, 26
Burke Act, 39, 64
Burke, Charles, 98–99

Carter, Charles D., 56, 60, 106
Chandler, O.K., 53, 98–99
Cherokee
    citizenship, 17, 30
    conflict within, 15–16
    intermarriage, 47, 96
    land use, 12
    relations with blacks, 46–47, 96
    relations with Osage, 87
    sovereignty, 16
    unity, 91–92, 96
Chickasaw
    blood, 31, 94
    citizenship, 18, 93–94
    grafters, 58
    intermarriage, 47–48
    relations with whites, 18
    taxation, 94
    unity, 92
Chickasaw Protective Association, 93, 94
Choctaw
    ancestry, 31
    blood, 31
    citizenship, 17, 30
    conflict within, 16, 18–19
    emigration, 23
    freedmen, 46
    graft, 58
    politics, 58
    removal, 3
    Snakes, 26
    sovereignty, 16
Chronicles of Oklahoma, 3
Civil War, 14–15, 20, 26, 29, 60
Collier, John, 101–102, 109–110

149

conservatives
    blood, 30, 32, 34, 41
    concern with whites, 17–18
    conflict with progressives, 25, 32–33, 72
    culture, 5, 11
    Dawes Commission, 6, 20–21
    desire to retain sovereignty, 21
    real Indians, 34–35, 37, 69
    relations with federal government, 61–62
    separatists and separatism, 6, 24–25, 27, 32, 35, 44, 52, 69, 88
Cornelius, C.P., 91, 96
Crazy Snake, *see* Harjo, Chitto
Creek
    allotment, 32
    conflict within, 18–19
    graft, 58
    land use, 12
    relations with blacks, 46, 49
    Snakes, 25, 28, 30, 32, 38, 43–44

Dawes Commission, 6, 20–21, 24–25, 29, 32, 40, 48, 93
Dawes, Henry, 20, 73, 104
Dawes reforms, 23–25, 30, 90, 97
Debo, Angie, 4
Dodge, Mabel, 101

Eaton, Rachel Caroline, 87, 100
Education, 60, 74–75

Fabens, Charles H., 81, 83
Fall, Albert, 72–73, 101
federal government
    blood, 40–41
    competency, 39–40, 61, 64–66, 74
    district agents, 60–63
    exploitative, 73, 79, 93
    forced competency of, 64–68
    freedmen, 48
    protective of Five Tribes, 6, 38, 53, 60–63, 65, 7, 79, 97, 98, 100
    reforms, 73–74
    trust restrictions, 38–40, 59, 60, 66–67, 74, 93
Five Tribes
    blood, 31, 40, 88–89, 98, 102–105, 107, 110
    citizenship requirements, 17, 30, 46
    conflict within, 14–15, 18–19, 21, 35, 72, 107, 110
    constitutional government, 13
    diversity within, 10–11, 29–30
    education, 12, 74–75, 97, 100
    emigration, 23–24, 34
    exploitation of, 81–84
    freedmen, 46–47
    intermarriage, 42, 47
    oil, 79–81
    population, 11
    powerlessness, 6
    racial mingling, 23–24, 47, 49
    relations with whites, 16–17
    solidarity/unity, 6–7, 72, 84–85, 95, 99–100, 104, 107, 109–110
    sovereignty, 10, 13–15, 19, 49, 50
    within Oklahoma, 6, 17, 20, 37, 43, 45
Four Mothers Society, 28

General Federation of Women's Clubs, 81, 101
grafters, 51, 57–59, 61, 65–68, 71–72, 77–83, 89, 97
Green Corn Rebellion, 75
Green Peach War, 18–19
Gritts, Levi, 89, 91–92
guardians, 27

Harjo, Chitto, 25–26, 28, 30–31, 43–44, 48

Index

Harkins, George, 3–5
Hickory Ground, 25

identity, 45
Indianness, 5, 29, 95–96, 105
    blood, 5, 23–24, 30, 33–35,
        49–50, 88–89, 96, 105, 107
Indian Reorganization Act, 109–110
Indian Rights Association, 81, 101
Indian School Journal, 60–61
Indian Territory, 23
    geography, 11

Jackson, Jacob, 23
Johnston, Douglas, 92

Keetoowah Society
    allotment, 27
    Cherokee tribe, 63, 89–92
    elections, 26, 27
    Night Hawks 89–91
    race, 30
    relations with government, 26,
        28, 62, 89–91, 98–99
    relations with other tribes, 28
    unity, 89–90

Land Rush, 20
Locke, Victor, Jr., 75, 85, 92
Locke-Jones War, 18–19

McCumber, 39
McPherson, O.M., 66
Mexico, 24
Mott, M.L., 63

oil, 6, 72, 75–84
Oklahoma
    economic change, 54–57, 76,
        84
    Ku Klux Klan, 75–76, 78
    socialists, 55, 75
    taxation, 59, 94
    World War I, 55–56, 64, 72,
        75–76

Osage, 77–78, 97
    oil, 77–78, 101
    violence against, 77–78
Oskison, John, 103–106
Owen, Robert, 39, 61

Parker, Gabriel, 64–69, 72
Porter, Pleasant, 9–10, 21, 25, 32
Posey, Alex, 26, 33, 74
Progressives
    assimilationist, 5, 21, 37–38,
        41, 64–65, 69, 88, 97
    blood, 41, 105
    culture, 12, 33
    disassociating with blacks, 6,
        38, 44–45, 48–49
    disassociating with full bloods,
        38, 42–43, 69, 95
    education, 74–75
    land use, 2
    relations with full bloods,
        69–71
    relations with whites, 18,
        37–38, 42
    slavery, 12, 46

racial thinking, 29, 30
Rogers, W.C., 26–27, 63

Seeley Chapel Association, 92
Seeley Indian Association, 93
Sells, Cato, 65
Seminoles
    allotment, 21
    conflict within, 26
    factionalism, 94
    relations with blacks, 47, 49,
        94–95
    unity, 94
Sequoyah, 104
Smith, Redbird, 89–90
Sniffen, Matthew K., 81, 83
Society of Oklahoma Indians,
    95–97, 99

Valentine, Robert, 64–65

Waite, Stand, 100

Williams, Robert L., 67–68
World War I
    effect on Five Tribes, 64, 68, 89–90
    Selective Service, 68
Wright, Muriel, 3–5

PICKENS COUNTY LIBRARY
110 WEST FIRST AVE.
EASLEY, SC 29640
    191 0735